nF419858

Instant Agriculture Success

The Author

 **Atul Kumar Singh,** working as Rural Agriculture Extension Officer in Department of Agriculture (C.G.). He has also worked as Agriculture expert level-1 in Kisan Call Centre, Jabalpur, Field Executive in BASIX Micro-finance, Dewas and Block Coordinator in ATMA Project. He has received B.Sc. (Ag.) and M.Sc. (Ag.) degrees in the field of Extension Education from Jawaharlal Nehru Krishi Vishwa Vidyalaya, Jabalpur (M.P.) and Ph.D from Mahatma Gandhi Chitrakoot Gramoday Vishwavidyalaya, Chitrakoot (M.P.). He has qualified National Eligibility Test (ASRB, ICAR) examination. He has authored book "Agriculture Extension Explorer", many research papers in journals of repute. He has vast experience of teaching, research and extension activities.

Instant Agriculture Success

(for all Agricultural Competitive Examination)

Useful for

- *AIEEA-ICAR-JRF/NTS (PGS), SRF, NET & ARS Examinations*
- *Ph.D. Entrance Examinations of IARI, IGKV, BHU, GBPUAT & various Indian Agricultural Universities*
- *ADA, ADO, RAEO, ASCO & Bank AO Examinations*

Atul Kumar Singh

2018

Daya Publishing House®

A Division of

Astral International Pvt. Ltd.

New Delhi – 110 002

ISBN 9789390384099 (PB)

Publisher's Note:

Every possible effort has been made to ensure that the information contained in this book is accurate at the time of going to press, and the publisher and author cannot accept responsibility for any errors or omissions, however caused. No responsibility for loss or damage occasioned to any person acting, or refraining from action, as a result of the material in this publication can be accepted by the editor, the publisher or the author. The Publisher is not associated with any product or vendor mentioned in the book. The contents of this work are intended to further general scientific research, understanding and discussion only. Readers should consult with a specialist where appropriate.

Every effort has been made to trace the owners of copyright material used in this book, if any. The author and the publisher will be grateful for any omission brought to their notice for acknowledgement in the future editions of the book.

Published by   :   **Daya Publishing House®**
A Division of
Astral International Pvt. Ltd.
– ISO 9001:2015 Certified Company –
4736/23, Ansari Road, Darya Ganj
New Delhi-110 002
Ph. 011-43549197, 23278134
E-mail: info@astralint.com
Website: www.astralint.com

Digitally Printed at   :   **Replika Press Pvt. Ltd.**

Preface

Agriculture in India plays a major role in economic development. Agriculture is back bone of Indian economy. Agriculture in India is means of livelihood of almost two third of the work force in the country. The Indian Council of Agricultural Research (ICAR), New Delhi is conducting all India competitive examination every year for awarding JRF, SRF, NET, ARS *etc*. All these examination are mostly objective based and students always look for study material that is ready to use and easy to grasp. There are only few books available in market on agriculture which completely satisfy the requirement of the students particularly for those who are preparing for competitive examinations. Keeping in view, it was felt to bring out a book which could serve the basic and innovative knowledge of various fields of agriculture.

The present book "**Instant Agriculture Success**" has been prepared in most simple and appreciate manner. The main objective of this book is to help the reader to quickly grasp the facts comprehensively and systematically from various branch of agriculture (Agronomy, Meteorology, Horticulture, Genetics and Plant Breeding, Entomology, Pathology, Agriculture Extension, Agriculture Economics, Agriculture Statistics, Veterinary and Animal husbandry etc.). Above mentioned topics are not involved in any other single book but you will find whole above mentioned topics are available in this book.

Here, I would like to express my heartfelt thanks to the great persons who taught me to do something for others with the following rules. I came to get all what I have today I also thanks all who helped me to compile this book, especially to Dr. Baldeo Singh, Dr. S.R.K. Singh, Dr. V. K. Pyasi, Dr. Sanjay Singh, Mr. R. K. Sharma and my college senior and batchmates for his hard work to compile this book.

I am thankful to my wife Smt. Priyanka Singh who helped me for material collection and typing.

I dedicate this book to my grand father Shri R.D. Singh and whole family.

Readers are welcome to print out error and omissions. If any, and send there valuable suggestion for improving the quality of book.

Atul Kumar Singh

Contents

List of Co-Authors

Agronomy, Weed management	Dr. Mahendra Singh (Assistant Professor, B.H.U), Shailesh Pandey (Ph.D Scholar, MGCGVV), Rakesh Sharma (A.D.A, Govt. of CG)
Irrigation and Water Management, Meteorology, Environmental Science and Agro-forestry, Plant Physiology, Soil Science, Fertilizer, Biochemistry, Pathology, Mushroom Cultivation	Rakesh Sharma (A.D.A, Govt. of CG), Atul Kumar Singh (Ph.D Scholar, MGCGVV), Ashish Singh Baghel (Team Leader, IWMP)
Horticulture	Ashutosh Ghanekar (M.Sc. Horticulture), Devendra Upadhyay (Scientist, IGKV, Raipur) Abhishek Tiwari (R.A.E.O., Govt. of C.G.)
Genetics and Plant Breeding, Seed Technology, Biotechnology	Dr. Nidhi Pathak (SRF, JNKVV, Jabalpur), Vijay Katara (T.A., JNKVV, Jabalpur), Manish Kumar Mishra (A.D.A, Govt. of M.P.)
Entomology, Fisheries, Apiculture, Sericulture	Brishbhan Ahirwar (Ph.D Scholar, MGCGVV),

Agricultural Extension, About ICAR and Research Institute, Useful information about agriculture	Atul Kumar Singh (Ph.D Scholar, MGCGVV, Chitrakoot), Dr. Abhay Wankhede (Scientist, JNKVV), Dr. K.K.Singh (Associate Professor, MGCGVV), Dr. D.P. Rai (Associate Professor, MGCGVV), Dr. Arup K. Gupta (Associate Professor, MGCGVV), Rahul Mishra (BTM in ATMA, M.P. Govt.) Akhilesh Kulhare (Ph.D Scholar, MGCGVV),
Agricultural Economics	Dr. Ashutosh Shriwastav (Professor, JNKVV, Jabalpur) Prem Ratan Pandey (Ph.D Scholar, MGCGVV)
Agricultural Statistics	Dr. Umesh Thakur (Assistant Professor, JNKVV, Jabalpur)
Agricultural Engineering, Veterinary and Animal Husbandry	Atul Kumar Singh (Ph.D Scholar, MGCGVV), Rakesh Sharma (A.D.A, Govt. of CG)

1
Agronomy

☆ Agriculture word derived from — **Latin word**

☆ Agronomy word derived from — **Greek word**

☆ Father of Agronomy — **Pietro de Crescenzi**

☆ Writer of Horse-hoeing Husbandary — **Jethro Tull**

☆ Kharif crops are generally denotes as — **Short day plants**

☆ Rabi crops are generally denotes as — **Long day plants**

☆ Cotton, Maize, Sunflower, Safflower, Groundnut, Tomato, Buck wheat are — **Day neutral plants**

☆ Cereals are botanically — **Caryopsis**

☆ Recommended fertilizer dose (N:P:K) of cereal crops are — **4:2:1**

☆ Recommended fertilizer dose (N:P:K) of pulse crops are — **1:2:1or 1:2:2**

☆ Recommended fertilizer dose (N:P:K) of oilseed crops are — **3:2:1**

☆ Recommended fertilizer dose (N:P:K) of fodder and fibre crops are — **2:1:4**

☆ Paddy, Wheat, Gram, Pea, Mustard, Cotton, Barley, Arhar, Soybean are — **C_3 Plants**

☆ Maize, Sugarcane, Sorghum, Bajra are — **C_4 plants**

☆ Pineapple is — **CAM plant**

☆ Mechanization index is found highest in — **Wheat**

☆ Sowing pattern used in dry land — **Broadcasting**

- ☆ Wavelength longer than 750 m/μ is not visible to eye, and are called — **Infrared**

- ☆ Mechanical manipulation of soil is called — **Tillage**

- ☆ Tillage increase the water holding capacity and infiltration of soil but reduce — **Bulk density**

- ☆ One acre is equal to — **0.404 ha**

- ☆ Tillage operation mainly aims to break, open and turn soil is — **Primary tillage**

- ☆ Tillage operation done to create a good seedbed for proper planting — **Secondary tillage**

- ☆ Before harvesting of paddy, the Seed of succeeding crops is broadcast at 10 to 15 days is — **Utera crops**

- ☆ Sustainable agriculture can be achieved by — **More technological support**

- ☆ Objective of Sustainable agriculture is — **Maintain ecological balance**

- ☆ Which family of crops is more exhaustive — **Gramineae**

- ☆ Post harvest losses for cereals — **10 per cent**

- ☆ Extensively grown pulse crop in India — **Chickpea**

- ☆ First AICRP was started in which crop — **Maize 1957**

- ☆ Dropsy disease in human being is caused by which weed — *Argemone mexicana*

- ☆ Crop having maximum area under irrigation — **Wheat**

- ☆ Mostly crops are cultivated in — **Acidic nature**

- ☆ 0.1 per cent solution represents how much ml in 500 liters of water — **500 ml**

- ☆ Sugarcane and Paddy are — **Aminophilic plants**

- ☆ Plant rectangularity is — **Plant to Plant or Row to Row distance**

- ☆ Cultivation of two or more than two crops of different height simultaneously on a certain piece of land in a certain period is called — **Multistoried cropping**

- ☆ Land is fixed but crop is rotated year after year on same land is called — **Crop rotation**

- ☆ Slitting open of a pod or fruit in characteristics manner at maturity is known as — **Dehiscent (*e.g.* - pods of pulse)**

I. CULTIVATION PRACTICES OF AGRONOMICAL CROPS

Kharif Crops

Paddy

☆ Botanical name	– *Oryza sativa*
☆ Family	– **Poaceae (Gramineae)**
☆ Chromosome number (2n)	– **24**
☆ Origin	– **South East Asia**
☆ Inflorescence called	–**Panicle**
☆ Fruit type of paddy grain is	– **Caryopsis**
☆ Paddy is a	– **Short day plant**
☆ Pollination type	– **Self pollinated**
☆ Seed rate of paddy for SRI method	– **15 kg/ha**
☆ Seed rate of paddy for transplanting	– **50-60 kg/ha**
☆ Seed rate of paddy for broadcasting	– **80-100 kg/ha**
☆ Seed rate of paddy for dribbling	– **80-90 kg/ha**
☆ Father of hybrid paddy	– **Yuan Longping**
☆ Seed rate in Dapog method of nursery	–**1 kg/m²**
☆ Mat type nursery is related to	– **Paddy crop**
☆ Recommended fertilizer dose (N:P:K)	– **100:60:40 kg/ha**
☆ Recommended dose for Zn application in paady field	– **25 kg/ha**
☆ Correction Iron chlorosis in paddy	– **Spray 1 per cent solution of ferrous sulphate**
☆ Gaseous loss of N_2 fertilizer from soil is called	– **Denitrification**
☆ Optimum pH for cultivation	– **4-6**
☆ Optimum Temperature for good growth	– **30-32°C**
☆ Optimum Temperature for ripening	– **20-25°C**
☆ Spacing for short duration varieties	– **20 × 10 cm²**
☆ Spacing for medium duration varieties	– **20 × 15 cm²**
☆ Paddy Protein (Oryzein) content	– **7 per cent**
☆ Aroma in paddy is due to presence chemical of	– **Di-acetyl propaline**
☆ Nursery area required for paddy transplanting is	– **1/10th ha (1000 m²/ha)**

☆ Widely used nitrogenous fertilizer in paddy — **Ammonium sulphate**

☆ Which gas emitted from paddy field — **Methane (CH_4)**

☆ Paddy take nitrogen in form of — **Ammonical form (NH_4)**

☆ Maximum production of paddy in the world is — **China**

☆ 2nd Maximum production of paddy in the world is — **India**

☆ Maximum paddy exporter in the world is — **Thiland**

☆ Highest Productivity of paddy in the world is — **Japan**

☆ Highest production and productivity in India — **West Bengal**

☆ Gene responsible for dwarf varieties of paddy — **Dee-Gee-Woo-gene**

☆ Javanica paddy is grown in — **Indonesia**

☆ Wild type of paddy is — **Javanica (Intermediate)**

☆ Japonica paddy is grown in — **Japan**

☆ Paddy having short stem and no lodging tendency — **Japonica (Temperate)**

☆ Indica paddy is grown in — **India**

☆ Paddy having long stem and lodging tendency — **Indica (Tropical)**

☆ Paddy seeds are sown directly of the main field is — **Upland cultivation**

☆ Dapog method was introduced in India from — **(IRRI) Phillipines**

☆ Types of paddy cultivation — **Up land, Low land and Deep water**

☆ Paddy seeds are sown directly of main field is called — **Up land**

☆ Area required for planting under Dapog method of nursery — **25-30 m^2/ha**

☆ Seedlings ready for transplanting in Depog method — **12-14th days**

☆ Normally paddy plant transplanted at — **20-25 Days After Sowing (DAS)**

☆ In SRI method paddy transplanted at — **10-12th DAS**

☆ SRI method was originated in — **Madagascar (1983)**

☆ Aman paddy is also known as — ***Aghani/kharif***

☆ Sowing time of Aman paddy — **May - June**

☆ Harvesting time of Aman paddy — **October – December**

☆ Aus/Autumn paddy is also known as — ***Pre - Kharif***

☆ Sowing time of Aus/Autumn paddy — **March - April**

☆ Harvesting time of Aus/Autumn paddy — **August–September**

☆ Boro paddy is also known as — **Summer/Spring**

☆ Sowing time of Boro paddy — **December – January**

☆ Harvesting time Boro paddy — **April- May**

☆ Butachlor herbicide use as — **Pre-emergence**

☆ Most similar weeds of paddy — *Echinocloa colonum* and *Echinocloa crusgallli*

☆ Sanwa weed botanical name — *Echinocloa colonum*

☆ Common weedicide used in paddy field is — **Anilophos** and **Butachlor**

☆ Yield loss by weeds in paddy field — **10 per cent**

☆ Best biofertilizer for paddy — **Azolla**

☆ Nitrogen fixation bacteria found on root surface of paddy is — **Azospirillum**

☆ Neck blast damage is severe in — **Basmati varieties**

☆ The most important critical stage of paddy for irrigation — **Booting stage**

☆ For low land paddy fertilizer is applied in — **Reduced zone only**

☆ Nitrogen use efficiency in paddy is around — **30-50 per cent**

☆ White eye disease is caused by — **Fe deficiency**

☆ Khaira disease is caused by — **Zn deficiency**

☆ Akiochi disease is caused by — **H_2S toxicity**

☆ Dead heart and white ear of paddy is caused by — **Yellow stem borer**

☆ Important cultural practice in paddy field — **Puddling**

☆ parboiling of rice conserve the vitamin is — **B_{12}**

☆ Hydrothermal treatment of paddy grain before milling is called — **Parboiling**

☆ Hulling Percentage of paddy — **70-75 per cent**

☆ Paddy harvesting moisture percentage — **20-22 per cent**

☆ Paddy storage moisture percentage — **14 per cent**

☆ First develop dwarf variety of paddy — **Taichung Native (TN-1)**

☆ Jagannath is the mutant variety of — **T-141**

☆ Miracle paddy variety — **IR-8 and TN-1**

☆ Super paddy concept given by — **G. S. Khush**

☆ World's first Super paddy variety for Alkaline/ Saline soil is — **Lunishree**

☆ Super paddy variety is developed by — **CPRI**

☆ World first high yielding and semi dwarf Basmati variety is — **Pusa Basmati**

☆ First hybrid of basmati in world — **PRH-10**

☆ Blast resistant varieties — **IR64, Tulsi**

☆ Deep water paddy varieties — **Jagannath, Pankaj**

☆ Saline alkaline soil suitable varieties — **CSR-10, CSR-27, CSR-13**

☆ China country is pioneer in — **Hybrid paddy**

☆ Level of water is maintained in wet nursery of paddy — **5 cm**

☆ Optimum depth of puddling in paddy — **5 cm**

☆ How much water is consumed for production of 1 kg of paddy — **5000 litre of water**

☆ Khaira disease first reported by — **Y.L. Nene at Pantnagar (1966)**

☆ Killer disease of paddy — **Bacterial leaf blight (BLB) and Tungro virus**

Maize

☆ Botanical name — ***Zea mays***

☆ Family — **Poaceae (Gramineae)**

☆ Origin — **Mexico**

☆ Inflorescence called — **Tessel**

☆ Also called — **Queen of cereals**

☆ Maize is a — **C_4 Plant**

☆ Style is very long filament, cluster of which is known as — **Silk**

☆ Quality protein maize (QPM) varieties released by using — **Opaqua-2 genes**

☆ Maize Protein (Zein) content — **8-10 per cent**

☆ Oil percentage — **4-5 per cent**

☆ Seed rate of hybrid maize — **20-25 kg/ha**

☆ Seed rate of composite maize — **15-20 kg/ha**

☆ Seed rate for fodder — **40-60 kg/ha**

☆ Seed rate in *kharif* — **15-20 kg/ha**

☆ Suitable temperature for maize growth — **30-32°C**

☆ Most critical stage for irrigation in maize — **Tasseling to Silking stage**

☆ Critical stages of maize for fertilizer — **Germination, Knee high, Tasseling**

☆ Water requirement for maize — **500-800 MM**

☆ Plant first appearance is — **Tassels**

☆ Leaves show red and purple colour due to defiency of — **Phosphorus**

☆ Maize variety widely grown in India — ***Zea mays indurate* (Flint corn)**

☆ Maize variety widely grown in USA — ***Zea mays identata* (Dent corn)**

☆ Sweetest maize spicies — ***Zea mays saccharata* (Sweet corn)**

☆ Maize species produce starch similar to tapioca — ***Zea mays ceretina* (Waxy corn)**

☆ Pop corn maize — ***Zea mays everta***

☆ Pop corn maize varieties — **Pearl pop corn, Amber**

☆ Fodder maize crop varieties — **African tall, J1006**

☆ QPM varieties — **HQPM, Shakti, Shaktiman1, Shaktiman2**

☆ Hybrid varieties — **Ganga-1, Ganga-4, Ranjit, Sangam, Ganga safed, Deccan-107**

☆ Composite varieties — **Vikram, Kishan, Sona, Vijay, Prabhat, Pratap, Pusa comp-1**

☆ Opaque -2 composite varieties — **Rattan, Protina, Shakti**

☆ Opaque -2 composite varieties are rich essential amino acids — **Lysine**

☆ Sweet maize variety — **African tall**

☆ Sweet corn variety — **Priya**

☆ Baby corn variety — **Prakash, VL-42**

☆ Commom herbicide use in maize — **Simazine**

☆ Maize harvesting moisture percentage — **20-25 per cent**

☆ Leading state of rabi season maize growing — **Bihar**

☆ First hybrid maize released in — **1961**

☆ Single cross technique of maize is given by — **E. M. East and G. H. Shull (1910)**

☆ Double cross technique of maize is given by — **D.F. Jones (1920)**

☆ 1st state in area and production in India — **U.P.**

☆ All India Coordinated Maize Improvement Project started in — **1957**

Sorghum (Jowar)

☆ Botanical name — *Sorghum vulgare (old - Sorghum bicolour)*

☆ Family — **Poaceae (Gramineae)**

☆ Chromosome number (2n) — **20**

☆ Origin — **Africa**

☆ Also called — **Camel crop**

☆ Inflorescence called — **Panicle**

☆ Seed rate — **12-15 kg/ha**

☆ Jowar is poor in lysine but rich in — **Leucine**

☆ First hybrid variety of sorghum — **CSH-1 (1965)**

☆ Grain and fodder purpose varieties — **CSH-13, CSV-15**

☆ Combine Kafir-60 is a — **Male sterile variety**

☆ Alkaloid content present in sorghum leaves — **HCN (Dhurin alkaloid synthesize in roots)**

☆ Major pest — **Shoot fly, Midge, Stem borer**

Pearl Millet (Bajra)

☆ Botanical name — *Pennisetum glaucum*

☆ Family — **Poaceae (Gramineae)**

☆ Chromosome number (2n) — **14**

☆ Origin — **Africa**

☆ Also known as — **Bulrush millet**

☆ Protein per cent — **10-12 per cent**

☆ Seed rate — **5 kg/ha**

☆ Seed rate for fodder — **20-30 kg/ha**

☆ India is the first country to develop and use of
hybrid of pearl millet — **HB-1 (1965)**

☆ Highest productivity in India — **U.P.**

☆ 80 per cent phosphorus in bajra grains stored in the form of — **Phytate**

☆ Bajra is a sensitive to — **Acidic soil and water logging condition**

Groundnut (Peanut/Earth Nut/Monkey Nut/Manillanut)

☆ Botanical name — *Arachis hypogea*

☆ Bunchy/Erect/Spanish type groundnut — *Arachis hypogea* **var.** *fastigate*

☆ Spreading/Verginia runner type groundnut — *Arachis hypogea* **var.** *procumbens*

☆ Family — **Leguminosae**

☆ Origin — **Brazil**

☆ Protein content in groundnut — **26 per cent**

☆ Protein content in shell — **7 per cent**

☆ Oil percentage — **40-45 per cent**

☆ Fruit type — **Nut**

☆ Groundnut is — **Modified fruit**

☆ Nitrogen percentage in groundnut cake — **7-8 per cent**

☆ Bitterness of groundnut kernel is due to — **Affalotoxin**

☆ Best soil for cultivation of groundnut — **Sandy loam soil**

☆ Seed rate for bunch type — **100-120 kg/ha**

☆ Seed rate for spreading type — **80-100 kg/ha**

☆ Spacing for bunch type — **30×10 cm²**

☆ Spacing for spreading type — **45×10 cm²**

☆ Sowing time — **15 June – 30 July**

☆ Bunching type varieties — **Jyoti, AK-12,Junagarh-11, Kishan**

☆ Spreading type varieties — **Chandra, Vikram, Verginia, TMV-1,Gangapuri**

☆ Fertilizer dose N:P:K — **20:40:20 kg/ha**

☆ Gypsum use for cultivation — **400 kg/ha**

☆ Important critical stage for irrigation — **Flowering, Pegging, Pod formation**

- ☆ Most suitable irrigation method — **Check basin method**
- ☆ Earthing-up is done in — **35-45 DAS**
- ☆ Chemical used for floral initiation — **NAA(40 ppm on 40 DAS)**
- ☆ Pegging stage — **55 DAS**
- ☆ Interculture practice avoided in groundnut at — **Pegging stage**
- ☆ Gynophore of groundnut is known as — **Peg**
- ☆ Bacteria used for N_2 fixation by nodule formation — ***Rhizobium japonicum***
- ☆ Which type groundnut shows dormancy — **Spreading type**
- ☆ Chemical used to break dormancy for spreading type groundnut — **GA₃**
- ☆ Which groundnut germinate in field before harvest — **Bunch type**
- ☆ Which chemical used to break dormancy for bunch type groundnut — **Malic hydrazide (MH)**
- ☆ Early leaf spot — ***Cercospora arachidicola***
- ☆ Late leaf spot — ***Cercospora personata***
- ☆ Rostee disease is due to — **Virus**
- ☆ Vector of virus in groundnut is — **Aphid**
- ☆ Which fungi affect kernel during storage — ***Aspergillus flavus***
- ☆ High yielding type of groundnut is — **Spreading type (Late maturity)**
- ☆ Highest producer state in India — **Gujarat**
- ☆ Shelling percentage — **70 per cent**
- ☆ Major pest — **White grub**
- ☆ Major disease — **Tikka (leaf spot)**

Sesamum (Til)

- ☆ Botanical name — ***Sesamum indicum***
- ☆ Family — **Pedaliaceae**
- ☆ Seed rate for line sowing — **3-4 kg/ha**
- ☆ Seed rate for broad casting — **5-7 kg/ha**

Green gram (Moong)

- ☆ Botanical name — ***Phasiolus aureus* (Old – *Vigna radiata*)**
- ☆ Family — **Leguminosae**
- ☆ Seed rate — **12-15 kg/ha**

☆ Yellow mosaic resistant variety — **Basanti, Sumrat, Pant moong-3**

☆ Early maturing variety — **Pusa baisakhi, PS-16, K-851, SML-668**

☆ Protein percentage — **25 per cent**

Black Gram (Urad)

☆ Botanical name — *Phasiolus mungo* (Old – *Vigna mungo*)

☆ Family — **Leguminosae**

☆ Seed rate — **20-25 kg/ha**

☆ Spacing(R-R *P-P) — **40×10 cm²**

☆ Varieties — **Barkha, Type-9 (T-9), Gwalior-2, JU-2,Pant U-30, Pant U-19**

Soybean

☆ Botanical name — *Glycine max*

☆ Family — **Leguminosae**

☆ Seed rate — **75 kg/ha**

☆ Depth of sowing — **3 Cm**

☆ Popularly known as — **Wonder crop**

☆ Soybean designated as — **Boneless meat**

☆ Protein content in soybean — **40-42 per cent**

☆ Oil content in soybean — **20-22 per cent**

☆ Highest production state in India — **M.P.**

☆ Most common cultivated soybean in India — **Yellow coloured soybean**

☆ Bacteria used for N_2 fixation by nodule formation — *Rhizobium japonicum*

☆ N_2 fixation per hectare by soybean — **35-40 kg/ha**

☆ Varieties — **JS-335, JS-93-05, PK-472, Ankur, Gaurav Clark, JS-2**

Cotton

☆ Botanical name — *Gossypium* spp.

☆ Family — **Malvaceae**

☆ Indian/old world cotton — *Gossypium arborium, Gossypium herbaceum*

☆ American/new world cotton — *Gossypium hirsutum*

☆ Egyptian/sudan/sea island cotton — *Gossypium barbadense*

☆ Chromosome number (2n) Indian cotton — 26

☆ Chromosome number (2n) American cotton — 52

☆ Origin — India

☆ Seed rate of American cotton — 12-15 kg/ha

☆ Optimum plant population for Bt cotton — 10000 Plants/ha

☆ Oil content — 15-25 per cent

☆ Fibre colour of American cotton — Creamy White

☆ Cotton is popular in America as — White gold

☆ Bad opening of boll is called — Tirak

☆ Maturity of cotton fibre judge by — Arealometer

☆ One bale of cotton is equal to — 170 kg

☆ Ginning percentage — 24-43 per cent

☆ Largest producer state in India — Gujarat

☆ Chemical used for delinting of cotton — H_2SO_4

☆ Which part of the cotton plant contains lint and fuzz — Hemp

☆ The first commercial cotton variety of the world — Hybrid -4

☆ Hybrid-4 variety developed by — Dr. C. T. Patel (1970)

☆ Interspecific hybrid varieties of cotton — HB-224, Varalaxmi, DHB-105

☆ Indian cotton variety — G-777

☆ MSP by govt. for cotton variety of — F414, H-777, H4

☆ Fibre of cotton contains — Cellulose

☆ Seed cotton contain — 1/3 per cent lint and 2/3 per cent cotton seed (Binola)

☆ Leaves of cotton becomes radish colour due to — Decrease nutrient uptake and moisture

☆ Average weight of very fine fibre of cotton — <3.0 mg

☆ Less number of knots in cotton is called as — Superior quality cotton

☆ Formula for calculating of ginning per cent — Wt. of lint/wt. of seed cotton × 100

☆ Seed cotton is — Seed +Lint

☆ Ratio of cotton seed to lint is — 2:1

☆ Cotton seed is — Seed after removing lint

☆ Topping done at   – 80-90 DAS

☆ Bt cotton is resistant to   – **Spotted boll worm (Heliocoverpa)**

Jute

☆ Botanical name   – *Corchorus capsularis*

☆ Family   – **Tiliaceae**

☆ Seed rate   – **8-10 kg/ha**

☆ One bale is equal to   – **180 kg**

☆ Low quality of jute fibre attributed to   – **Discolouration of fibre**

☆ Retting of jute fibre is a   – **Biochemical process**

☆ Ideal stage of jute harvesting for fibre purpose   – **Small pod stage (135-140 DAS)**

☆ Varieties   – **JRC-7447(Shyamli), JRC-321 (Sonali), JRC-212 (Sabuj sona)**

Sugarcane

☆ Botanical name   – *Saccharum officinarum*

☆ Wild type   – *Saccharum spontaneum*

☆ Noble type/Tropical cane   – *Saccharum officinarum*

☆ Indian cane   – *Saccharum barberi* **and** *Saccharum sinense*

☆ Family   – **Gramineae**

☆ Origin   – **India**

☆ Best temperature for growth   – **21-27°C**

☆ Inflorescence is called   – **Arrow**

☆ Arrow stage arrives   – **300-350 days after planting**

☆ Seed rate   – **120000-125000 setts/ha for 1 budded, 45000-50000 setts/ha for 2 budded, 25000-30000 setts/ha for 3 budded setts**

☆ Chemical used for setts treatment   – **Agallal and Areton**

☆ Spacing (R-R)   – **90 cm**

☆ Fertilizer dose NPK   – **270:150:120 kg/ha**

☆ Planting material used for planting   – **Upper 1/3 to 1/2 part of cane**

☆ Upper 1/3 part is use used for sowing due to   – **High N_2 substance and glucose for better germination.**

☆ High dose of nitrogen decrease – **Sucrose content**

☆ Red rot resistant variety – **CO-19, CO-1148**

☆ Sets root are – **Temporary root**

☆ Shoot root are – **Permanent root**

☆ Flat bed method of sugarcane planting is most common in – **North India**

☆ Ridge and furrow method of sugarcane planting is most common in – **South India**

☆ Most critical stage for irrigation – **Formative stage (60-130 days after planting)**

☆ Adsali sugarcane planting time – **June-July (Kharif season)**

☆ Adsali sugarcane harvesting duration – **18 Months**

☆ Adsali sugarcane is common in – **Maharashtra**

☆ Eksali sugarcane planting time North India – **Feb.-April**

☆ Eksali sugarcane harvesting duration – **12 Months**

☆ Eksali sugarcane planting time South India – **Dec.-Jan.**

☆ Earthing up done at – **120 days after planting**

☆ Earthing should be done in the month – **June –July**

☆ Blind hoeind is recommended for – **Sugarcane**

☆ Sugarcane sowing in trench method to prevent – **Lodging**

☆ Tying should be done in the month – **August**

☆ Bacteria used for N2 fixation – ***Acetobacter diazotrophicus***

☆ Most common herbicide use – **Atrazine, Simazine, Alachlor**

☆ Major weeds – ***Cynodan dactylon* and *Sorghum halepense***

☆ Noble type sugarcane can be used for – **Chewing purpose**

☆ World largest sugarcane producer – **India**

☆ Highest production state in India – **U.P.**

☆ Highest productivity state in India – **Tamil Nadu**

☆ Highest sugar mills state in India – **U.P.**

☆ Highest producer of sugar from unit area – **Maharashtra**

☆ Nutrient responsible for translocation of sugar in sugarcane – **Potassium**

- ☆ Method of plant analysis for assessing nutrient requirement in sugarcane — **Crop logging**
- ☆ Sugar yield from sugar cane — **6-10 per cent from juice**
- ☆ Sucrose content in cane — **13-24 per cent**
- ☆ Jaggery extracted from juice is — **9-10 per cent**
- ☆ Criteria for harvesting of sugarcane — **Withering of lower leaves, brix 20 per cent, sucrose 15 per cent**
- ☆ Brix reading of juice indicates — **Total soluble solids (TSS)**
- ☆ Sugarcane is considered as mature when brix reading is — **18-20 per cent**
- ☆ Measuring maturity of sugarcane by — **Brix meter**
- ☆ By-product of sugarcane — **Molasses and Bagasses**
- ☆ Wonder cane is — **COC – 671 (highest sugar percentage)**
- ☆ Chemical used for ripening — **Glyphosate (5 kg/ha)**
- ☆ Burning of cane is done for — **Improve sucrose and juice quality**
- ☆ Sugarcane breeding institute (SBI) — **Coimbatore (T.N.)**
- ☆ Indian institute of sugarcane research (IISR) — **Lucknow (U.P.)**
- ☆ Indian sugar institute (ISI) — **Kanpur (U.P.)**
- ☆ Major disease of sugar cane is — **Red rot**

Pigeon Pea (Red gram/Arhar)

- ☆ Botanical name — *Cajnus cajan*
- ☆ Early maturing — *Cajnus cajan flavus*
- ☆ Late maturing — *Cajnus cajan bicolor*
- ☆ Family — **Leguminosae/Papilionaceae**
- ☆ Origin — **South Africa**
- ☆ Protein content — **25 per cent**
- ☆ Seed rate — **12-15 kg/ha**
- ☆ Spacing for extra early variety — **50×30 cm^2**
- ☆ Spacing for early variety — **75×30 cm^2**
- ☆ Spacing for late variety — **90×30 cm^2**
- ☆ Temperature for germination — **30-35°C**

☆ Temperature for growth   – 20-25°C

☆ Type of germination   – Hypogeal

☆ World first hybrid variety of arhar is   – ICPH-8 (released by ICRISAT)

☆ Varieties   – Prabhat, Pusa ageti, Pusa-84

☆ Extra short duration variety   – UPAS-120

☆ Wilt resistant variety   – Mukta, Amar, Azad

☆ Zn defiency in arhar is rectified by spraying of   – 5 kg $ZnSO_4$ + 2.5 kg Lime/ha

☆ Harvesting index (HI)   – 19 per cent (lowest among in pulses)

☆ Highest production state in India   – U.P.

☆ Highest productivity state in India   – Bihar

Important Points of other Kharif Crops

☆ Botanical name of Sunhemp/Banaras hemp/ Bombay hemp   – *Crotalaria juncea*

☆ Family of Sunhemp   – Leguminosae

☆ Botanical name of Ragi/Finger miller   – *Eleusine coracana*

☆ Family of Ragi   – Gramineae

☆ Chromosome number (2n) of Ragi   – 36

☆ Botanical name of Kodo   – *Paspulum scrobiculatum*

☆ Family of kodo   – Gramineae

☆ Seed rate of kodo   – 6-8 kg/ha

☆ Chromosome number (2n) of kodo   – 40

☆ Botanical name of Cheena/Proso millet   – *Panicum millaceum*

☆ Family of cheena   – Gramineae

☆ Chromosome number (2n) of cheena   – 36

☆ Botanical name of Kakun/Foxtail   – *Setaria italia*

☆ Family of kakun   – Gramineae

☆ Chromosome number (2n) of kakun   – 18

☆ Coarsest of all food grain is   – Kodo

☆ Botanical name of Cluster bean/Gaur   – *Cyamopsis tetragonoloba*

☆ Family of gaur   – Leguminosae

☆ Millets belongs to   – C_4 plant

☆ Higher productivity among the millet — Finger millet

☆ Seed rate of grass pea — 40-50 kg/ha

☆ Seed rate of castoe — 10 kg/ha

☆ Seed rate of sugar beet — 8-10 kg/ha

☆ Seed rate of potato — 1000-1500 kg/ha

Rabi Crops

Wheat

☆ Botanical name — *Triticum aestivum*

☆ Family — **Poaceae (Gramineae)**

☆ Maxican dwarf wheat — *Triticum aestivum* **(2n=42)**

☆ Bread wheat — *Triticum vulgare* **(2n=42)**

☆ Marconi/Duram wheat — *Triticum durum* **(2n=28)**

☆ Emmer wheat — *Triticum dicoccum* **(2n=28)**

☆ Indian dwarf wheat — *Triticum spherococum* **(2n=28)**

☆ Club wheat — *Triticum compactum*

☆ Origin — **South West Asia**

☆ Fruit type — **Caryopsis**

☆ Central zig-zag axis of wheat grain is called — **Rachis**

☆ Inflorescence called — **Ear**

☆ Flowring portion of wheat — **Ear/Head/Spike**

☆ Wheat protein (Gluten) content — **8-12 per cent**

☆ Starch in wheat grain — **60-70 per cent**

☆ Seed rate for timely sowing — **100 kg/ha**

☆ Seed rate for sowing by dibbler method — **25-30 kg/ha**

☆ When wheat seed is dropped by hand in furrow is called — **Kera Method**

☆ When wheat seed is dropped by nai of plough in furrow is called — **Pora method**

☆ If late sowing of wheat how much seed rate increase — **25 per cent**

☆ Spacing (R-R) — **22.5 cm**

☆ Depth of sowing dwarf varieties of wheat — **5 cm**

☆ Sowing depth of wheat depend on — **Length of coleoptile**

☆ Sowing time — 15th Nov to 20th Nov

☆ Word staple food grain is — Wheat

☆ For good bread quality which protein is essential — Gluten

☆ Recommended fertilizer dose (N:P:K) — 120:60:40 kg/ha

☆ Important sequence for irrigation stages — CRI, Hoot (late jointing), Milking

☆ Test weight — 40 g

☆ Gene responsible for dwarf varieties of wheat — Norin 10

☆ Triticum aestivum is a — Hexaploid plant

☆ Common bread wheat — *Triticum aestivum*

☆ Highest grown wheat species in India — *Triticum aestivum*

☆ Most critical stage for irrigation in wheat — Crown root initiation (CRI) 20-25 DAS

☆ Zinc and sulphates deficiency in wheat field reported in — Punjab

☆ Maximum area of irrigation wheat is in — Haryana

☆ Highest production state in India — U.P.

☆ Highest productivity state in India — Punjab

☆ Non traditional area for cultivating wheat is — Eastern India

☆ Important mimicry weed of wheat — *Phalaris minor* (Wild oat)

☆ Objectionable weed of wheat — *Convolvulus arvensis*

☆ Phalaris minor can be controlled by — Isoproturon or 2,4-D

☆ Common herbicide used to control weeds in wheat — 2,4-D

☆ Triple gene dwarf wheat varieties were released in — 1970

☆ Triticum aestivum was first introduced in India by — N. E. Borlaug

☆ Norin-10 gene was brought to USA by — S.C. Salamon (1948)

☆ Most of the present days India wheat varieties contain the gene — Rht1, Rht2

☆ First man made cereal is — Triticale

☆ Triticale is cross of — Wheat×Rye

☆ Most important wheat variety during green revolution — HD-2329

☆ Shelling percentage of wheat — 60 per cent

☆ Wheat moisture content at harvesting stage — 25-30 per cent

☆ Grain :straw ratio in maxican wheat — 1:1.5

☆ FIRB method has been evolved for sowing of — **Wheat**

☆ Suitable for late sown varieties — **Sonalika**

☆ Mutant variety of wheat — **Sabarmathi sonora**

☆ Suitable for rainfed area — **Sujata, C-306, Mukta, Shera**

☆ Single gene dwarf varieties — **Girija, Sonalika, UP-262, WL-711**

☆ Double gene dwarf varieties — **Arjun, Pratap, Kalyansona, UP-215, HD-2204**

☆ Triple gene dwarf varieties — **Hira, Moti, Jyoti, Sangam, Jawahar, UP-301, UP-319**

☆ Suitable for normal sown varieties — **Lermarojo, Kalyansona, Sonoro64**

☆ Associated weed of wheat — ***Phalaris minor, Chenopodium album, Avena fatua***

☆ Initial distinguishing character for identification of *Phalaris minor* — **Basal node is pink upto 50 DAS**

☆ Brown rust/Leaf rust Causal organism — ***Puccinia recondite***

☆ Black rust/Stem rust Causal organism — ***Puccinia graminis tritici***

☆ Yellow rust/Stripe rust Causal organism — ***Puccinia striformis***

☆ Wheat storage moisture percentage — **10 per cent**

Critical Stages

1. Crown Root Initiation (CRI) – 21 Day after sowing (DAS)
2. Tillering stage – 40 - 45 DAS
3. Jointing stage – 60 - 65 DAS
4. Flowering stage – 80 - 85 DAS
5. Milking stage – 100 - 105 DAS
6. Dough stage – 115 - 120 DAS

Barley

☆ Botanical name of Barley — ***Hordeum vulgare***

☆ Botanical name Two rowed Barley — ***Hordeum distichum*/H. *irregular***

☆ Botanical name Six rowed Barley — ***Hordeum vulgare***

☆ Family — **Gramineae**

☆ Seed rate — **75-80 kg/ha**

☆ Most critical stage for irrigation　　　　　– **Active Tiplering Stage (30-35 DAS)**

☆ Pearl barley is suitable for　　　　　　　　　– **Kidney disorders**

☆ Grassy weed in barley field can be controlled by　– **Isoproturon and 2,4-D**

☆ Lugri is a fermented drink developed from　　– **Hull less barley grains**

☆ High melting quality variety　　　　　　　　– **Rekha**

☆ Salt resistant variety of barley　　　　　　　– **Ratna**

☆ Molya disease resistant variety　　　　　　　– **RD-2052**

☆ Highly salt tolerance cereal crop　　　　　　– **Barley**

Chick Pea (Bengal gram/Gram)

☆ Botanical name　　　　　　　　　　　　　– *Cicer arietinum*

☆ Desi/Brown gram　　　　　　　　　　　　– *Cicer aeritinum*

☆ Kabuli/White gram　　　　　　　　　　　– *Cicer kabulium*

☆ Family　　　　　　　　　　　　　　　　– **Leguminosae**

☆ Chromosome no. (2n)　　　　　　　　　　– **16**

☆ Origin　　　　　　　　　　　　　　　　– **South West Asia**

☆ King of pulse　　　　　　　　　　　　　– **Gram**

☆ Optimum time for sowing　　　　　　– **15th -20th October**

☆ Seed rate　　　　　　　　　　　　　　– **75-80 kg/ha**

☆ Spacing　　　　　　　　　　　　　　– **30×10 cm²**

☆ Depth of sowing　　　　　　　　　　　– **7-10 cm**

☆ Sour taste of gram leaf is due to　　– **Malic acid and Oxalic acid**

☆ Fruit type　　　　　　　　　　　　　　– **Pod**

☆ Most frost affected crop among all field crops　– **Gram**

☆ Protein content　　　　　　　　　　　　– **21 per cent**

☆ Best soil for cultivation　　　　　　　　– **Light alluvial soil**

☆ Root system of gram　　　　　　　　　　– **Tap root system**

☆ Pollination type　　　　　　　　　　　– **Self pollination**

☆ Critical stage for irrigation　– **Pre flowering, pod developing stage**

☆ Wilt resistant variety　　– **JG-74, JG-315, Awarodhi, BG-256**

☆ Drought resistant variety　　　　　　　　– **NP-58**

☆ Best variety for dry land area　　　　　　– **C-235**

☆ Early maturing variety — JG-62, Chaff

☆ Suitable for rainfed condition variety — **Vishal, Anubhav, Gaurav**

☆ Avrodhi variety of gram is resistant to — **Wilt**

☆ Late planting of gram is done to protect the seedlings from — **Wilt disease**

☆ Removal of apical buds (30-40 DAS) to encourage lateral branching of gram is called — **Nipping**

Field Pea

☆ Botanical name — *Pisum sativum* **var.** *arvense*

☆ Garden pea — *Pisum sativum* **var.** *hartense*

☆ Family — **Leguminosae**

☆ Chromosome number (2n) — **14**

☆ Seed rate — **75 kg/ha**

☆ Spacing for field pea (R-R * P-P) — **30×5 cm²**

☆ Use of field pea — **Pulse purpose**

☆ Use of garden pea — **Green pods used for vegetable**

☆ Bacteria used for N_2 fixation — *Rhizobium leguminosarum*

☆ Leaf less variety of field pea — **Arpana**

☆ Fertilizer dose NPK — **20:50:30 kg/ha**

☆ Varieties of field pea — **Ambika, Hans, Rachana, T-65, T-163, KP-885**

☆ Shelling percentage — **49 per cent**

Lentil

☆ Botanical name — *Lens esculenta*

☆ Family — **Leguminosae**

☆ Seed rate — **30-40 kg/ha**

Mustard

☆ Botanical name — *Brassica* **spp.**

☆ Botanical name of Indian/Brown mustard — *Brassica juncea*

☆ Botanical name of sarson — *Brassica compestris*

☆ Family — **Crucifereae**

☆ Seed rate — **4-6 kg/ha**

☆ Fruit of mustard is known — **Siliqua**

☆ Which acid is present in mustard and rapseed — **Erusic acid**

☆ Critical stage for irrigation — **Rostte stage and siliqua formation stage**

☆ Varieties of brown mustard — **Pusa kalyani, BSH-1, Sufla**

☆ Varieties of mustard — **Krishna, Pusa bold, Kranti, Varuna, Vardan, Rohni**

☆ Hybrid variety of pusa jai kisan is also called — **Bio-902**

☆ Storage moisture percentage — **7-8 per cent**

Linseed/Flax

☆ Botanical name — *Linum usitatissimum*

☆ Family — **Linaceae**

☆ Seed rate — **25-30 kg/ha**

☆ Oil content — **40-42 per cent**

☆ Linolinic acid present in linseed oil — **50-60 per cent**

☆ Process of treatment of stalks for final fibre extraction is termed as — **Retting**

☆ Fertilizer dose NPK — **60:40:20 kg/ha**

☆ Varieties — **Kiran, Mukta, Sweta, Jawahar-7, Gourav, Shital**

Sunflower

☆ Botanical name — *Helianthus annus*

☆ Family — **Compositae**

☆ Seed rate — **5-7.5 kg/ha**

☆ Planting space — **50×20 cm^2**

☆ Also known as — **Non-conventional oilseed crop**

☆ High quality edible oil content — **45-50 per cent**

☆ PUFA content is highest in — **Sunflower**

☆ Head of sunflower is called — **Capitulai**

☆ Crop used as Protective crop is — **Sunflower**

☆ Non-conventional oilseed crop — **Sunflower**

☆ Rancidity in sunflower oil due to — **Oxidation**

☆ Varieties — **Jwalamukhi, MSFH-8, Modern, JS-1**

Safflower

☆	Botanical name	– *Carthamus tinctorius*
☆	Family	– **Compositae**
☆	Seed rate	– **15-20 kg/ha**
☆	Also known as	– **Fencing/Border crop**
☆	Fruit of safflower is called	– **Achene**
☆	Oil content	– **32-36 per cent**
☆	Varieties	– **JSH-129, JSI-7, EB-7, JSF-1**

Oat

☆	Botanical name	– *Avena sativa*
☆	Family	– **Gramineae**
☆	Seed rate	– **80-90 kg/ha**
☆	Best stage for harvesting	– **Dough stage**
☆	Variety	– **UPO-50, Afterlee, Kent, Algerian, Craig, Fleming gold, HFO-114**

Berseem/Eqyptian clover

☆	Botanical name	– *Trifolium alexandrinum*
☆	Family	– **Leguminosae**
☆	Seed rate	– **25-30 kg/ha**
☆	Varieties	– **Chindwara, IGFRI-99, Vardan, BL-1, Maskavi (C-10)**
☆	Sowing of berseem crop is done by	– **Broadcastring**
☆	Bacteria used for N_2 fixation by nodule formation	– *Rhizobium trifolium*
☆	First cutting done at	– **50-55 DAS**
☆	Objectionable weed is	– *Chicorium intybus* (Kasini)

Tobacco

☆	Botanical name	– *Nicotiana tabacum*
☆	Family	–**Solanaceae**
☆	Seed rate	– **2.5-3 kg/ha**
☆	Transplanting time	– **7-9 weeks (4-5 leaf stage)**
☆	Most critical stage for irrigation	– **Topping**

☆ Nicotine content accumulates from which part — **Leaves**

☆ Priming method of harvesting is popular in — **Cigarette, Wrapper and Chewing type**

☆ Fuel curing is done for — **Cigarette tobacco**

☆ Fire curing is done for — **Bidi, snuff, chewing, hookah tobacco**

☆ Removal of lateral branches or auxillary bud is called — **De-suckering**

☆ Curing is a — **Dry process**

☆ *Nicotiana tabacum* is growing for purpose of — **Smoking and chewing**

☆ Nicotiana rustica is growing for purpose of — **Hookah, chewing and snuff**

☆ Nicotine percentage in *Nicotiana tabacum* is — **0.5-5.5**

☆ Nicotine percentage in *Nicotiana rustica* is — **3.5-8.0**

☆ Central Tobacco Research Institute (CTRI) situated in — **Rajmundari (A.P.)**

☆ Bidi Tobacco Research Station situated at — **Anand (Gujarat)**

☆ Cigarette tobacco is prominent growing state is — **Andhra Pradesh and Karnataka**

☆ Mutant variety — **Bhavya, Jayashri**

Lucerne/Alfalfa

☆ Botanical name — *Medicago sativa*

☆ Family — **Leguminosae**

☆ Seed rate — **20-25 kg/ha**

☆ Bacteria used for N_2 fixation by nodule formation — *Rhizobium meliloti*

☆ Stem parasitic weed — *Cuscuta reflexa* **(Doddar)**

☆ Varieties — **Anand-2, Type-9, IGFRI-5, Moopa, Rambler**

☆ Lucern yellowing disorder caused by — **Boron**

Cow Pea

☆ Botanical name — *Vigna unguiculata (old – Vigna sinensis)*

☆ Family — **Leguminosae**

☆ Seed rate — **20-25 kg/ha**

☆ Also known as — **Vegetable meat**

Important Points of other Rabi Crops

☆	Pseudo cereals is	– **Buck wheat**
☆	Agrostology is the study of	– **Grasses**
☆	Most prominent and adopted cropping pattern in India	– **Paddy - Wheat**
☆	Cropping intensity of maize-potato-wheat is	– **300 per cent**
☆	Moong/Urad + maize is an example of	– **Parallel cropping**
☆	Traditional method of cultivation in hilly area is called	– **Jhoom farming**
☆	Jhoom cultivation mostly found in	– **Eastern part of India**
☆	Richest source of protein among the food grain	– **Pulse**
☆	Directorate of pulse research	– **Kanpur (U.P.)**
☆	Which pulse crop does not fix nitrogen from atmosphere	– **Rajma**
☆	Which oil is extract from endosperm	– **Coconut oil**
☆	Which oil is suitable for heart patient	– **Sunflower oil**
☆	Most widely grown rabi pulse crop	– **Bengal gram**
☆	Maximum oil seed crop produced in India	– **Groundnut**
☆	Pulse word derived from	– **French word**
☆	Legume term derived from	– **Latin word**
☆	Hydroponics term derived from	– **Greek word**
☆	Hydroponics is also known as	– **Soilless culture**
☆	Passive sub-irrigation is also known as	– **Semi hydroponics**
☆	In Wheat, mustard may be grown as inter crop in the row ratio of	– **9 Wheat : 1 Mustard**
☆	Land Equivalent Ratio (LER) is used to evaluate	– **Intercropping**
☆	LEISA is associated to	– **Organic farming**
☆	Seed rate of Niger	– **5-8 kg/ha**
☆	Rhizobium, Azotobacter, Azospirillum, Acetobactor, Azolla, BGA are	– **Nitrogen fixer**
☆	Bacillus, Pseudomonas, Aspergillus, Penicillium are	– **P Solubiliser**
☆	Example of erosion permitting crop	– **Maize**
☆	Example of erosion resistant crop	– **Cow pea**
☆	100 g of Azolla contains	– **0.5 g N, 0.5 g P, 0.4 g Ca, 0.5 g Mg, 0.5g Fe**

☆ Cropping intensity formula – = **Total cropped area/Net sown area × 100**

☆ Rotational intensity formula – = **No. of crops grown in rotation/ Duration of the rotation ×100**

☆ Cropping index formula – = **No. of crops grown per year/ Cropped area ×100**

2
Irrigation and Water Management

☆ Concept of permanemt wilting point (PWP) proposed
by   **– Briggs and Shantz (1912)**

☆ Concept of available water given by   **– Veihmayer and Hendrickson (1981)**

☆ Concept of potential evapotranspiration (PET)
given by   **– Thornthwaite (1948)**

☆ Water harvesting term was first used by   **– Myors**

☆ Artificial application of water to supply moisture essential
to plant growth is called   **– Irrigation**

☆ Situation where all the pores both micro and macro
are filled with water is called   **– Water logging**

☆ A natural hydrological unit having common runoff
outlet point is called   **– Watershed**

☆ Collecting and storage of water on the surface of soil
for subsequent use   **– Water harvesting**

☆ Horizontal flow of water in irrigation channels is known as   **– Seepage**

☆ Vertical or downward movement of water from
different soil layer is called   **– Percolation**

☆ Flow of excess water from the field after saturation of
soil is called   **– Run - off**

- ☆ First entry of water from the upper layer of soil is known as — **Infiltration**
- ☆ Infiltration occurs in — **Unsaturated soil**
- ☆ Downword movement of nutrients and salts from the root zone with the water is called — **Leaching**
- ☆ In which period irrigation is supplied to a crop is — **Base period**
- ☆ Biggest river basin of India is — **Ganga**
- ☆ Discharge rate of water in sprinkler irrigation system — **More than 1000 lit/ hr at 2.5 Bar or 2.5 kg/cm^2 Pressure**
- ☆ Discharge rate of water in drip irrigation system — **1-4 litres/hrs at 2.5 Bar or 2.5 kg/cm^2 Pressure**
- ☆ Drip irrigation system is discovered at — **Israel**
- ☆ Country having highest irrigation efficiency — **Israel**
- ☆ Country having highest irrigation area — **India**
- ☆ Water harvesting in situ is known as — **Runoff farming**
- ☆ Growth retardant type antitranspirents is — **Cycocel**
- ☆ Reflectant type antitranspirents is — **Kaoline (5 per cent)**
- ☆ Simplest way of adoption of plant to drought is — **Evasion**
- ☆ Chemical accumulated during drought condition — **Proline**
- ☆ Which is accumulate in the leaves of water stressed plants — **ABA**
- ☆ Which irrigation method has highest irrigation efficiency — **Drip method**
- ☆ Water requirement for paddy is — **90-250 cm**
- ☆ Water requirement for Cotton is — **70-130 cm**
- ☆ Water requirement for Sugarcane is — **150-250 cm**
- ☆ Water requirement for Wheat, sorghum, tobacco, soybean is — **45-65 cm**
- ☆ Agronomical measures are adopt only where land slop is — **< 2 per cent**
- ☆ Mechanical measure are adopted by where land slop is — **> 2 per cent**
- ☆ In water logged soil, high concentration of — **Methane**
- ☆ Drainage of one ha cm in 24 hrs is equal to — **1.16 litre/sec**
- ☆ Most popular mechanical measure to control soil erosion and conserve is — **Contour bunding**
- ☆ The crop grown on degraded land for improvement is called — **Conservation crop**

☆ Most efficient method of irrigation is — **Drip irrigation**

☆ Drip irrigation method is also known as — **Trickle irrigation**

☆ Vertical mulches are used only in — **Black cotton soil**

☆ Crops growing for conserve soil moisture known as — **Mulch crops**

☆ Which type of water is most available to plants — **Capillary water**

☆ Capillary water is also known as — **Available water**

☆ Water held up to the tension about 31 atm is known as — **Hygroscopic coefficient**

☆ Irrigation method suitable for low land paddy and jute — **Flooding method**

☆ Irrigation method which is suitable for saline soils — **Flood method**

☆ Most common method of surface irrigation to irrigate groundnut and pulse — **Check basin**

☆ Suitable method of irrigation to irrigate fruit trees — **Ring basin**

☆ Lines connecting the points of equal water table are called — **Isobath**

☆ One hectare centimetre of irrigation is equal to — **100000 litres**

☆ One hectare meter of irrigation is equal to — **10000 litres**

☆ One cu feet of water is equal to — **28.32 litres**

☆ Running water is measured by — **Cusec**

☆ Irrigation method suitable for undulated land, sandy soil, vegetable and fruit crops — **Sprinkler method**

☆ Which method is suitable for wider spaced orchard crops, sugarcane and for saline soils — **Drip method**

☆ Saving of water in sprinkler and drip irrigation as compare to surface irrigation methods — **25-50 per cent and 50-75 per cent**

☆ A diffusive process by which liquid water in the form of vapour is lost in atmosphere — **Evaporation**

☆ Under water logged conditions which nutrients are found deficient for the crops — **Zn and Cu**

☆ The root developed due to water logging in most of the crops — **Adventitious root**

☆ Cultivation of crops in areas where average annual rainfall is <750 mm — **Dry farming**

☆ Cultivation of crops in areas where average annual rainfall is 750 to 1150 mm — **Dry land farming**

☆ Cultivation of crops in areas where average annual rainfall is >1150 mm — **Rainfed farming**

☆ Crop growing season of dry land farming is — **75-120 Days**

☆ Change in normal crop planning to meet weather abnormalities is termed as — **Contingent planning**

☆ Chemical used to check transpiration losses of water — **Antitranspirants**

☆ Stomatal closing type antitranspirents is — **2,4-D, Atrazine and PMA at low concentration**

☆ Total depth of water required by a crop is called — **Delta**

☆ Area irrigated by one cusec discharge of water is — **Duty of water**

☆ Alternate crops recommended to sown under late onset of monsoon — **Green gram, Black gram, Sunflower, Cow pea, Castor**

☆ According to USDA classification, the land belongs to class VI and VII are suitable for — **Timber cum fiber farming**

☆ Stages of water erosion are — **Splash → Sheet → Rill → Gully → Ravine**

☆ Stages of wind erosion are — **Saltation → Suspension → Surface creep**

☆ Micro-watershed covers an area of about — **100-1000 ha (catchments command area)**

☆ Major irrigation project covers an area of — **>10000 ha (catchments command area)**

☆ Medium irrigation project covers an area of — **2000 - 10000 ha (catchments command area)**

☆ Minor irrigation project covers an area of — **<2000 ha (catchments command area)**

☆ Life saving irrigation is also known as — **Contingency irrigation**

☆ 75 per cent of rain fall is received by — **S-W monsoon (June to September)**

☆ Soil moisture is determined by — **Tensiometer**

☆ Water flow is Measurement by — **Parsall flume**

☆ Film farming type antitranspirents is — **Hexadeconal, Wax, Silicon and Mobileaf**

☆ Which crop rotation under dry land situation will be more remunerative — **Sesamum-Gram**

☆ Crop sown under condition of early onset of monsoon — **Sesamum, Pearl millet**

☆ Most appropriate crops in dry land farming — **Sorghum, Gram, Pearl millet, Toria**

☆ Full form of LEISA — **Low External Input Sustainable Agriculture**

☆ Contour bunding is adopted where — **Land slope (6 per cent) and in areas average annual rain fall is < 600 mm**

☆ Bench terracing is practiced on — **Steep slopping (16-33 per cent) and undulated land**

☆ Detachment and transportation of top soil particles by wind or water is known as — **Soil erosion**

☆ Types of soil movement in the process of wind erosion — **Saltation, Suspension, Surface creep**

☆ About 50-75 per cent of soil erosion by wind is carried out by — **Saltation**

☆ Very fine soil particle (<0.1 mm diameter) eroded by mechanism — **Suspension**

☆ Removal of soil particles due to rain drops is called — **Splash erosion**

☆ Which mechanism of water erosion is known as "Death of Farmers" — **Sheet erosion**

☆ Chanalization begins from which mechanism of water erosion — **Rill erosion**

☆ The advantage stage of gully erosion is — **Ravines**

☆ The land capability classes suitable for crop cultivation are — **Class I to III**

☆ Tension at field capacity ranges from — **-0.1 to -0.33**

☆ permanemt wilting point (PWP) utilized dwarf sunflower as an — **Indicator plant**

☆ permanemt wilting point (PWP) soil moisture tension — **-15 bars**

☆ Portion of capillary water lying between field capacity (1/3 atm) and wilting coefficient (15 atm) is — **Available water**

☆ Water retained by soil in capillary pores against gravity (1/3 to 31 bar) by the force of surface tension — **Capillary water**

☆ Supplemental irrigation is known as — **Life saving Irrigation**

3
Weed Management

☆ Weed term was firstly used by — **Jethro Tull**

☆ Unwanted plant, growing where it is not desired — **Weed**

☆ Off type crop varieties are — **Rogue**

☆ Undesirable or off type or unwanted plant either weeds or another variety of same crop are called —**Rogue plants**

☆ Unwanted plant are removed from seed production field before flowering is called — **Rouging**

☆ Such weeds, that are grown in cultivated field — **Obligate weeds**

☆ Cropped along with wild land weed are known as — **Facultative weeds**

☆ Weed that depends for its growth on its host plant — **Parasitic weed**

☆ Undesirable, troublesome weed difficult to control — **Noxious weed**

☆ Serious weed of Wheat — *Phalaris minor*

☆ Alachlor, butachlorand propanil belongs to which chemical group — **Amide group**

☆ Atrazine, Simazine belongs to which chemical group — **Triazines group**

☆ Pendimethalin, Fluchloralin belongs to which chemical group — **Dinitroanilines group**

☆ Trade name of Chlorimuron ethyl — **Kloben**

☆ Trade name of Glyphosate — **Round up**

☆ Trade name of Nitrofen — **Toke E-25**

☆ Trade name of Pendimethalin — **Stomp**

☆ Trade name of Alachlor — **Lasso**

☆ Trade name of Fenoxa prop-ethyl — **Whip super**

☆ Trade name of Ethoxy sulfuron — **Sunrise**

☆ Trade name of Chlorimuron 10 per cent + Metasulfuron methyl 0 per cent — **Almix**

☆ Chemical name of Basalin is — **Fluchloralin**

☆ Chemical name of stomp is — **Pendimethalin**

☆ Herbicide are not used in dust formulation because of — **Drifting hazards**

☆ Water hyacinth, Hydrilla, Salvania are example of — **Aquatic weed**

☆ Most dominant aquatic weed *Eichhornia crassipes* is controlled by — *Neochetina bruchi*

☆ Which weed having herbicide resistance — *Avena fatua*

☆ Which is a indicator plant for the bioassay of Atrazine — **Soybean**

☆ First commercial bio-herbicide is — **DEVINE**

☆ BIPOLARIS is used to control weed for — **Johnson grass**

☆ *Cyperus rotundus* is a example of — **Absolute weed**

☆ Paddy in wheat field is a example of — **Relative weed**

☆ *Chenopodium album* is a example of — **Obligate weed**

☆ Parthenium is a example of — **Noxious weed**

☆ Phalaris minor in wheat field is a example of — **Mimicry weeds**

☆ Wild paddy in paddy field is a example of — **Mimicry weeds**

☆ Best example of sedge is — *Cyperus rotundus*

☆ First biological controlled weed is — *Lantana camara*

☆ *Lantana camara* is also used as — **Ornamental plant**

☆ *Parthenium hysterophorus* is biologically controlled by — *Zygrogramma bicolarata*

☆ Which cause more wastage of herbicide by drift — **Ultra low volume spray**

☆ Bright red coloured triangle in herbicide shows — **Extremely toxic group**

☆ 2-4D belongs to which chemical group — **Chloro - phenoxy compound**

☆ Glyphosate, Anilophos belongs to which chemical group — **Organophosphorus group**

☆ A herbicide that kill only target plants on weeds while crops are not affected is called — **Selective herbicide**

☆ 2,4-D, Atrazine, Simazine, Butachlor and Fluchloralin belongs to — **Selective herbicide group**

☆ a herbicide that kill all vegetation that they come in contact is called — **Non-selective herbicide**

☆ Diquat, Paraquat, Glyphosate, Oxadiargyl belongs to — **Non-selective herbicides group**

☆ A herbicide that kills plants when they come in contact with plants is called — **Contact herbicide**

☆ Paraquate, Diquat are example of — **Contact herbicide**

☆ Paraquate, Diquat is associated to the group of — **Bipyridiliums**

☆ Propanil is a — **Contact selective herbicide**

☆ Which herbicide shows knock down effect — **Diquate, Paraquate and Glyphosate**

☆ A herbicide that move within the plant to effect as herbicide is called — **Systemic herbicide**

☆ 2,4-D, Atrazine, Propanil are example of — **Systemic herbicide**

☆ Paraquate, Diquat, 2,4-D are example of — **Pre-emergence herbicide**

☆ Fluchloralin is example of — **Pre-planting incorporation herbicide**

☆ Diuron, Atrazine, Methyl bromide are — **Soil sterilant**

☆ Effective herbicides on mono cotyledons weeds are — **Delapon, Fluchloalin**

☆ Herbicide which have low residual toxicity — **Diquat, Paraquat**

☆ Herbicide which have high residual toxicity — **Diuron, Atrazine**

☆ Stale seed bed technique of weed control is — **Cultural method**

☆ Which stage of crop are more prone to weed competition — **Germination to seedling**

☆ Critical period of crop weed competition for transplanted paddy — **30-45 DAS**

☆ Herbicide applied 1 day before sowing/planting are comes under — **Pre-plant incorporate herbicide**

☆ Alachlor, Fluchloralin, Trifluralin are — **Pre-plant incorporate herbicide**

☆ Herbicide applied 1-3 days after sowing (DAS) are comes under — **Pre-emergence herbicides**

☆ Alachlor, Atrazine, Butachlor, Nirofen, Pendimethalin, Simazine are — **Pre-emergence herbicides**

☆ Herbicide applied 25-40 DAS comes under — **Post- emergence herbicides**

☆ 2-4D, Chlorimuron ethyl, Diquat, Fenoxaprop-ethyl, Isoproturon, Paraquat, Sulfosulfuron are — **Post- emergence herbicides**

☆ Problematic weed, whose seed once mixed with crop seed is extremely difficult to separate is called — **Objectionable weed**

☆ 2,4-D used as a herbicide for — **Broad leaved weed**

☆ Practice of flushing out germinable weed seeds before crop sowing is called — **Stale seed bed**

☆ One plant having detrimental effect on other plants by releasing root chemical through roots — **Allelopathy**

☆ A weed that complete their life cycle in one year is called — **Annual weeds**

☆ Phalaris minor, Echinocloa colonum, Amaranthus are example of — **Annual weeds**

☆ A weed that complete their life cycle in five years is called — **Binneal weeds**

☆ Alternanithra echinate, Eichorrutim intybus are example of — **Binneal weeds**

☆ A weed that complete their life cycle in more than two years is called — **Perennial weeds**

☆ Cynodon dactylon, Cyprus rotundus are example of — **Perennial weeds**

☆ Bathua, Krishna neel, Satyanashi, Hiran khuri are — **Rabi weed**

☆ Striga is semi root parasite found in — **Maize, Sorghum, Sugarcane**

☆ Herbicide used in zero tillage — **Paraquat and Diquat**

☆ Directorate of weed science situated at — **Jabalpur**

☆ Old name of Directorate of weed science is — **National Research Centre for Weed Science**

☆ Orobanche is total root parasite found in — **Solanaceous crop**

☆ Total root parasite — **Orabanchi (Associated with Tobacco)**

☆ Semi root parasite — **Striga (Associated with Sorghum)**

☆ Total stem parasite — **Cuscuta (Associated with Lucerne crop)**

☆ Semi stem parasite — **Loranthus (Associated with fruit crops)**

☆ Orobanche is also known as — **Broom rape**

☆ Which type of photosynthesis possess by most of weeds — **C_4 type**

☆ Example of C$_4$ weed is — **Common lamsquarter**

☆ Chicory is an objectionable weed in — **Berseem**

☆ Chermical safeners also known as herbicide antidote *e.g.* — **NA (first safeners in maize)**

☆ The ability of plant to survive and reproduce after treatment with a dose of herbicides that would normally kill the plant is called — **Herbicide resistant**

☆ Weed biotype that has gained resistant to more than one herbicides with same mode of action is called — **Cross resistant**

☆ Weed biotype that has developed tolerance to more than one herbicides with different mode of action is called — **Multiple resistant**

Short Description of the 11 Most Commonly Used Herbicide Modes of Action

ACCase Inhibitors

Group	Chemical Family	Trade Names	Active Ingredient
1	Arloxyphenoxypropionate "FOPs"	Assure II	quizalofop
		Hoelon^r	diclofop
		Fusilade	fluazifop
		Puma	fenoxaprop
1	Cyclohexanedione "DIMs"	Select, Select Max, others	clethodim
		Poast, Poast Plus	sethoxydim
1	Phenylpyrazoline "DENs"	Axial XL	pinoxaden

ALS Inhibitors

Group	Chemical Family	Trade Names	Active Ingredient
2	Imidazolinone "IMIs"	Beyond, Raptor	imazamox
		Cadre	imazapic
		Pursuit	imazethapyr
		Scepter	imazaquin
2	Sulfonylurea "SUs"	Accent	nicosulfuron
		Ally	metsulfuron
		Amber	triasulfuron
		Autumn	iodosulfuron
		Beacon	primisulfuron
		Classic	chlorlumuron
		Express	tribenuron

Group	Chemical Family	Trade Names	Active Ingredient
		Glean	chlorsulfuron
		Harmony	thifensulfuron
		Maverick	sulfosulfuron
		Option	foramsulfuron
		Osprey	mesosulfuron
		Peak	prosulfuron
		Permit	halosulfuron
		Resolve	rimsulfuron
2	Triazolopyrimidine	FirstRate	cloransulam-methyl
		PowerFlex	pyroxsulam
		Python	flumetsulam
		Strongarm	diclosulam
2	Pyrimidinyl (thio)benzoate	Staple	pyrithiobac
2	Sulfonylaminocarbonyltriazolinones	Everest	flucarbazone
		Olympus	propoxycarbazone

Root Growth Inhibitors

Group	Chemical Family	Trade Names	Active Ingredient
3	Dinitroaniline	Treflan, others	trifluralin
		Prowl, others	pendimethalin
		Sonalan	ethafluralin

Growth Regulators

Group	Chemical Family	Trade Names	Active Ingredient
4	Phenoxy-carboxylic acid	Many	2,4-D
		Butyrac, others	2,4-DB
			MCPA
4	Benzoic acid	Banvel, Clarity, Status, others	dicamba
4	Pyridine carboxylic acid	Stinger	clopyralid
		Starane	fluroxypyr
		Tordon', Grazon'	picloram
4	Quinoline carboxylic acid	Paramount	quinclorac

Photosynthesis Inhibitors (Photosystem II)

Group	Chemical Family	Trade Names	Active Ingredient
5	Triazine	Aatrex[r], atrazine[r], others	atrazine
		Princep	simazine
		Caparol	prometryn
5	Triazinone	Sencor	metribuzin
		Velpar	hexazinone
5	Uracil	Sinbar	terbacil
6	Nitrile	Buctril, others	bromoxynil
6	Benzothiadiazinone	Basagran	bentazon
7	Urea	Linex, Lorox	linuron
		Karmex	diuron

Shoot Growth Inhibitors

Group	Chemical Family	Trade Names	Active Ingredient
8	Lipid synthesis inhibitor, thiocarbamate	Eptam	EPTC
15	Chloroacetamide	Dual, Cinch, others	metolachlor
		Intrro[r], Micro-Tech[r]	alachlor
		Harness[r], Degree[r], Surpass[r], others	acetochlor
		Outlook	dimethenamid-P
15	Oxyacetamide	Define	flufenacet

Aromatic Amino Acid Synthesis Inhibitors

Group	Chemical Family	Trade Names	Active Ingredient
9	Glycine	Roundup, Touchdown, others	glyphosate

Glutamine Synthesis Inhibitors

Group	Chemical Family	Trade Names	Active Ingredient
10	Phosphonic acid	Ignite, Liberty	glufosinate

Pigment Synthesis Inhibitors

Group	Chemical Family	Trade Names	Active Ingredient
12	Pyridazinone	Zorial Rapid 80	norflurazon
13	Isoxazolidinone	Command	clomazone

Group	Chemical Family	Trade Names	Active Ingredient
27	Triketone	Callisto	mesotrione
		Laudis	tembotrione
		Impact	topramezone
27	Isoxazole	Balance[r]	isoxaflutole

PPO Inhibitors

Group	Chemical Family	Trade Names	Active Ingredient
14	Diphenylether	Blazer	acifluorfen
		Reflex, Flexstar	fomesafen
		Cobra	lactofen
		Goal	oxyfluorfen
14	N-phenylphthalimide	Valor	flumioxazin
		Resource	flumiclorac
14	Thiadiazole	Cadet	fluthiacet
14	Triazolinone	Aim	carfentrazone
		Spartan, Authority	sulfentrazone

Photosynthesis Inhibitors (Photosystem I)

Group	Chemical Family	Trade Names	Active Ingredient
22	Bipyridilium	Gramoxone Inteon[r], others	paraquat
		Reglone, others	diquat

[r] Restricted use pesticide.

Critical period of crop weed competition and yield loss caused by weeds in different crops

Sl.No.	Crop	Critical period (DAS/DAT)	Yield Reduction (per cent)
(A)		Cereals	
(a)	Rice (Direct seeded)	15 – 45	15 – 90
(b)	Rice (Transplanted)	30 – 45	15 – 40
(c)	Maize	30 – 45	20 – 40
(d)	Sorghum	15 – 45	15 – 40
(e)	Pearl millet	30 – 45	15 – 60
(f)	Wheat	30 – 45	20 – 40
(B)		Pulses	
	Pigeonpea	15 – 60	20 – 40

Sl.No.	Crop	Critical period (DAS/DAT)	Yield Reduction (per cent)
	Green gram	15 – 30	25 – 50
	Black gram	15 – 30	30 – 50
	Cowpea	15 – 30	15 – 30
	Chickpea	30 – 60	15 – 25
	Pea	30 – 45	20 – 30
	Lentil	30 – 60	20 – 30
(C)	Oilseeds		
	Soybean	20 – 45	40 – 60
	Groundnut	30 – 60	40 – 50
	Sunflower	30 – 45	30 – 50
	Castor	30 – 60	30 – 45
	Sesamum	15 – 45	15 – 40
	Safflower	15 – 45	35 – 60
	Rapeseed mustard	15 – 40	15 – 30
	Linseed	20 – 45	30 – 40
(D)	Commercial crops		
	Sugarcane	30 – 120	20 – 30
	Potato	20 – 40	30 – 60
	Cotton	15 – 60	40 – 50
	Jute	30 – 45	50 – 80
(E)	Vegetable crops		
	Cauliflower	30 – 45	50 – 60
	Cabbage	30 – 45	50 – 60
	Okra	15 – 30	40 – 50
	Tomato	30 – 45	40 – 70
	Onion	30 – 75	60 – 70

The herbicides which have been found most potent for control of weeds in different crops

Crop	Herbicide	Rate of Application	Time of Application
	Butachlor	1.0 – 2.0 kg/ha	Pre-emergence
	Anilophos	0.45 kg/ha	Pre-emergence
Rice	2,4-D	0.50 kg/ha	Post emergence
(Direct seeded)	Fenoxaprop-p-ethyl	100 – 150 g/ha	Post emergence
	Pretilachlor	0.75 kg/ha + safner 750 g/ha	Pre-emergence

Crop	Herbicide	Rate of Application	Time of Application
Rice (Transplanted)	Butachlor	1.0 – 2.0 kg/ha	6 – 7 DAT
	Anilophos	0.45 kg/ha	6 – 7 DAT
	2,4-D	0.5 kg/ha	25 – 30 DAT
	Fenoxaprop-p-ethyl	100 – 150 g/ha	20 – 25 DAT
	Pretilachlor	0.75 kg/ha + safner 750 g/ha	Pre-emergence
Maize/Sorghum/ Bajra	Fluchloralin	0.75 – 1.0 kg/ha	PPI
	Pendimethalin	1.0 – 1.5 kg/ha	Pre-emergence
	Alachlor	1.5 – 2.0 kg/ha	Pre-emergence
	Metolachlor	1.0 – 1.5 kg/ha	Pre-emergence
	Metribuzin	0.5 kg/ha	Pre-emergence
	2,4-D	0.5 kg/ha	30 DAS
	Atrazine	2.0 kg/ha	Pre-emergence
Soybean	Alachlor	1.5 – 2.0 kg/ha	Pre-emergence
	Pendimethalin	0.75 – 1.0 kg/ha	Pre-emergence
	Metribuzin	0.35 – 0.5 kg/ha	Pre-emergence
	Quizolofop	50 g/ha	15 – 20 DAS
	Chlorimuron	8 – 12 g/ha	15 – 20 DAS
	Fenoxaprop (9EC)	80 – 120 g/ha	20 – 25 DAS
Soybean	Imazethapyr	75 – 100 g/ha	15 – 20 DAS
	Chlorimuron+Fenoxaprop	24 g +100 g/ha	15 – 20 DAS
	Oxidiargyl	100 g/ha	15 – 20 DAS
	Penoxsulam	20 g/ha	Pre-emergence
Groundnut	Alachlor	1.5 – 2.0 kg/ha	Pre-emergence
	Fluchloralin	0.75 – 1.0 kg/ha	PPI
	Pendimethalin	0.75 – 1.0 kg/ha	Pre-emergence
	Quizalofop	40 – 50 g/ha	15 – 20 DAS
	Imazethapyr	75 – 100 g/ha	15 – 20 DAS
Green gram/black gram	Pendimethalin	0.75 – 1.0 kg/ha	Pre-emergence
	Alachlor	1.5 – 2.0 kg/ha	Pre-emergence
	Quizalofop	40 – 50 g/ha	15 – 20 DAS
	Imazethapyr	75 – 100 g/ha	20 – 25 DAS
Cotton	Pendimethalin	0.75 – 1.0 kg/ha	Pre-emergence
	Oxadiazon	0.5 – 0.75 kg/ha	Pre-emergence
	Fluchloralin	0.75 – 1.0 kg/ha	PPI
	Diuron	0.5 – 0.75 kg/ha	Pre-emergence
	Trifloxysulfuron	10 – 20 g/ha + adjuvant	25 – 30 DAS

Crop	Herbicide	Rate of Application	Time of Application
Wheat	Sulfosulfuron	25 g/ha	25 – 30 DAS
	Carfentrazene	10 – 12 g/ha	20 – 25 DAS
	Carfentrazene + Isoproturon	20 g + 1.0 kg/ha	20 – 25 DAS
	Clodinofob + Isoproturon	60 g + 750 g/ha	20 – 25 DAS
	Clodinofop + carfentrazone	60 g +10 g/ha	20 – 25 DAS
	Fenoxaprop	100 g/ha	25 DAS

DAS= days after sowing; DAS= days after transplanting.

4

Meteorology

☆ Bioclimatic low proposed by — Hopkins

☆ Celsius and Fahrenheit relationship formula — C/5= F-32/9

☆ Mean distance between earth and sun — 1.5×10^8 km

☆ Temperature on the sun — 6000^0C

☆ Average percent of solar radiation reach to earth — 50 per cent

☆ Value of solar constant is — 1.94 cal/cm²/min

☆ Optical wave lengths used in remote sensing ranges from — 0.3 – 15 µm

☆ Which wind is responsible for rain fall in Tamil Nadu — North East Monsoon

☆ The radiation in the sunlight that gives us the feeling of hotness is — Infra red

☆ The radiation emitted by the sun and responsible for the cause of skin cancer — Ultra violet

☆ A weather condition over a given region during a longest period is called — Climate

☆ A condition of atmosphere at a given place and time is called — Weather

☆ Monsoon covers 85 per cent rain fall in India — South west monsoon

☆ North – East monsoon is also called — Retreating monsoon

☆ If the rain received 25 mm or more in a particular day is called — Rainy day

☆ Average size of rain drop is — 2 mm diameter

☆ Warm cloud seeding is done by use of — Sodium chloride (NaCl)

☆ Cold cloud seeding is done by use of chemical — Silver iodide (AgI_2)

☆ Phenomenon of warming of eastern pacific — *EI nino*

☆ Phenomenon of cooling of eastern pacific — *LI nino*

☆ Which surface has least Albedo — Moist black soil

☆ Atmosphere water is known as — Green water

☆ Soil water is known as — Blue water

☆ The unit used to record clouds — Okta

☆ Widely used index for classification of droughts — Palmer drought index

☆ Date of onset of mansoon in India — 1st June

☆ Date of mansoon withdrawal in India — 31st september

☆ Height of 8-18 km from earth surface is called — **Troposphere**

☆ Height of 18-50 km from earth surface is called — **Stratosphere**

☆ Height of 50-80 km from earth surface is called — **Mesosphere**

☆ Height of above 80 km from earth surface is called — **Ionosphere**

☆ Volume percentage of N_2 in atmosphere — **78.03 per cent**

☆ Volume percentage of O_2 in atmosphere — **20.95 per cent**

☆ Volume percentage of Ar in atmosphere — **0.93 per cent**

☆ Volume percentage of CO_2 in atmosphere — **0.03 per cent**

☆ Concentration of CO_2 in air — **0.03 per cent**

☆ Concentration of CO_2 in soil — **0.25 per cent**

☆ All weather phenomena like rain, fog, frost, cloud etc. occur in the zone of — **Troposphere**

☆ Closest and densest layer of atmosphere — **Troposphere**

☆ Ozone layer present in — **Stratosphere**

☆ Coldest region of the atmosphere — **Mesosphere**

☆ Radio transmission found in — **Ionosphere**

☆ Gas less zone — **Thermosphere**

☆ What is wave length of visible solar radiation — **0.39 - 0.70 mm**

☆ Wave length longer than 750 m/μ is not visible to eye is called — **Infra red**

☆ Wind turbine uses — **Kinetic energy**

☆ Heat flow in solid/soil by process of — **Conduction**

☆ Heat flow in liquid/water by process of — **Convection**

☆ Heat flow in air by process of — **Radiation**

☆ Carbon dioxide, Methane, Nitrous oxide is — **Green house gases**

☆ Chief green house gas responsible for global warming — **CO_2 (50 per cent)**

☆ Which green house gas linked with paddy crop — **Methane (CH_4)**

☆ Which gas responsible for ozone depletion — **CF_2Cl_2**

☆ Which gas is a substitute for CFCs — **Hydrofluorocarbons**

☆ Ozone depletion mainly due to — **Chloro Floro Carbon (CFCs)**

☆ Continuous temperature record by which instrument — **Thermo graph**

☆ Instrument used for measuring concentration of ozone in air — **Ozonometer**

☆ Low pressure area near the equator is called — **Doldrums**

☆ Lowest temperature in a day is observed at — **Just before sunshine**

☆ Total incoming solar radiation is measured by — **Pyranometer**

☆ Relative humidity is measured by — **Psychrometer**

☆ Evaporation is measured by — **Evaporimeter**

☆ Wind pressure/Strength is measured by — **Beaufort scale**

☆ Wind direction determined by — **Wind vane**

☆ Wind velocity is measured by — **Anemometer**

☆ Atmospheric pressure is measured by — **Barometer**

☆ Evapotranspiration is measured by — **Lysimeter**

☆ Metric suction is measured by — **Tensiometer**

☆ Rainfall is measured by — **Rain guage**

☆ Measurement of sun light intensity is expressed in — **Lux units**

☆ Shortest day in Northern Hemisphere — **21nd December**

☆ Longest day in Northern Hemisphere — **21st June**

☆ Duration of day and night are equal all over the earth — **20st March and 23 September**

☆ Jupiter planet is known as — **Lord of heavens**

☆ Indian standard time (IST) is calculated from a clock tower at the — **Allahabad observatory**

☆ Cyclones are generally originated in water warmer than — **27°C**

☆ Which instrument record temperature without contact the object — **Infrared thermometer**

☆ Nimbus word associated with — **Rain**

☆ Warm cloud seeding is done by — **Sodium chloride (NaCl)**

☆ Earth rotates in its axis at an angle of — **23 1/2°**

☆ Distance between Earth and Sun — **1.5×10^8 km**

☆ Photoperiod >12 hours — **Long day plants**

☆ Photoperiod < 12 hours — **Short day plants**

☆ Melting point of water — **32°F**

☆ Boiling point of water — **210°F**

☆ Solar constant is also known as — **Longley**

☆ Albedo also called — **Reflect radiation**

☆ Diameter of rainfall is — **0.5 mm**

☆ Diameter of drizzle is — **Less than 0.5 mm**

☆ Wind circulation in cyclone in northern hemisphere is — **Anti clock wise**

☆ Wind circulation in cyclone in southern hemisphere is — **Clock wise**

☆ Cyclone is also known as — **Typhoon**

☆ Cyclones are generally originated in water warmer than — **$27^{o}C$**

☆ First hint of cyclone is — **Assemblage of clouds**

☆ Average speed of cyclone is — **18 kmph**

☆ Centre of low pressure is called — **Cyclone**

☆ Centre of high pressure is called — **Anticyclone**

☆ Wind ranging from 60-120 kmph is called as — **Storm**

☆ Any structure which is reduce the wind speed is called as — **Wind break**

☆ Atmospheric pressure always decreases with — **Altitude**

☆ Ozone is — **Light blue gas**

☆ RH and Temperature is highest at — **Equator**

☆ At equator days and nights are always — **Equal**

☆ Atmospheric pressure is measured in — **Millibar**

☆ Wind is measured in — **Okta**

☆ Horizontal heat transfer from warmer area to cooler area — **Cloth line effect**

☆ Vertical heat transfer from warmer area to cooler area — **Oasis effect**

☆ Velocity of light in air — **3×10^{8} m/s**

☆ Solar constant value is — **1.94 cal/cm^2/min or 1353 Watts/m^2**

☆ Relative Humidity (R.H.) formula — **= Water vapour present in the air/Water vapour required for saturation ×100**

☆ Relative Humidity is expressed in — **Percentage (per cent)**

☆ Unit of absolute humidity — **g/m^3 or g/cc**

☆ Unit of specific humidity — **g/kg**

☆ Lines of equal temperature is called — **Isotherm**

☆ Lines of equal cloud cover is called — **Isonephs**

☆ Lines of equal atmospheric pressure is called — **Isobar**

☆ Lines of equal amount of rainfall is called — **Isohyets**

☆ Line of equal depth of rainfall is called — **Isopluvial**

☆ Lines of equal wind speed is called — **Isotach**

☆ Imaginary line connecting points of equal water elevation — **Isobaths**

☆ Line of equal sunshine hours — **Isohels**

☆ Line of equal evaporation value — **Isopleths**

☆ Period of very short range weather forecasting — **0-12 hrs**

☆ Period of short range weather forecasting — **< 3 days or Up to 72 hrs**

☆ Period of Medium range weather forecasting — **3 -10 days**

☆ Period of long range weather forecasting — **> 10 days**

☆ Forms of precipitation

    ☆ **Fog** – A thin cloud of varying size

    ☆ **Frost** – A frathery deposite of ice

    ☆ **Mist** – A very thin fog

    ☆ **Dew** – Moisture condensed in small drops upon cool surface

    ☆ **Rain** – Drop of >0.5 to 6 mm

    ☆ **Drizzle** – Drop of < 0.5 mm

    ☆ **Snow** – Ice crystal resulting from sublimation

5

Environmental Science and Agroforestry

☆ Forest word derived from   – **Latin**

☆ Ecosystem word was coined by   – **A.G. Tansley**

☆ Ecology term was introduced by   – **Hackel**

☆ Biodiversity term coined for first time by   – **E.O. Wilson**

☆ Indian forest Act was come in existence   – **1927**

☆ National wildlife protection Act was passed in   – **1972**

☆ Forest conservation Act was made in   – **1980**

☆ National forest policy was enunciated in   – **1988**

☆ Convention of biodiversity came into force on   – **29 December 1993**

☆ Indian institute of forest management (IIFM)   – **Bhopal**

☆ Forest school is established at   – **Dehradun**

☆ National Research Centre for Agroforestry is situated at   – **Jhansi (1988)**

☆ International Centre for Research in Agroforestry (ICRAF) is situated at   – **Nairobi (Kenya)**

☆ Van mahotsav day   – **1 July**

☆ Van mahotsav started by   – **K.M. Munshi**

☆ National biodiversity authority is established at — **Chennai**

☆ National biodiversity board is situated at — **New Delhi**

☆ National biodiversity authority was established in the year of — **2003**

☆ Biological diversity act was made in — **2002**

☆ United nation environment programme (UNEP) was launched in — **Nairobi (1972)**

☆ A system where agriculture and forestry are practised on the same unit of land either simultaneously or separately is called — **Agroforestry**

☆ Variety of living organisms (flora and fauna) is called as — **Biodiversity**

☆ Ecosystem component, which break down dead organic matter and wastes — **Decomposers**

☆ Which is not included under biomass — **Water**

☆ Ecosystem where producers are large size — **Grass land ecosystem**

☆ Which group of organisms convert light into food are called — **Autotrophs**

☆ Which organisms used to feed on producer or consumers are called — **Heterotrophs**

☆ Which organisms feed on plants and primary consumers are called — **Herbivores**

☆ Which organisms feed on meats and secondary consumers are called — **Carnivores**

☆ Which organisms feed on both plants and animals are called — **Omnivores**

☆ Plants which synthesize food them self through photosynthesis are called — **Phototrophs**

☆ Food chain base in the ocean is — **Phytoplankton**

☆ Primary consumer in aquatic system — **Zooplanktons**

☆ Plants which grow on other plants are called — **Epiphytes**

☆ Main constituent of CNG is — **Methane**

☆ Main constituent of LPG is — **Butane**

☆ Renewable natural resources are — **Water and Wood**

☆ Non renewable natural resources are — **Fuels, Fossil, Minerals**

☆ Ratio of height: width: length in shelterbelt system — **1:25:10 Meter**

☆ Global warming focuses on an increase in the level of which gas in the atmosphere — **CO_2**

☆ Colourless and odourless air pollutant is — SO_2

☆ Chief gas of green house — CO_2 and CH_4

☆ Most poisonous pollutant in water is — **Arsenic**

☆ Most commonly used disinfectant in water purification — **Chlorine**

☆ Permissible limit of iron in drinking water — **1ppm**

☆ Manimata disease is due to — **Mercury toxicity**

☆ Itai-Itai disease is due to — **Cadmium toxicity**

☆ Unit of energy is — **Joule**

☆ Most appropriate and effective type of crop cultivation in forests is — **Intercropping**

☆ Spacing maintained between hedge row intercropping in alley cropping — **4-8 Meter**

☆ Most suitable wood lot trees in India — **Casuarina and Leucaena**

☆ Most common example of taungya system — **Planting of teak in Myanmar**

☆ Nitrogen fixing tree — ***Leucaena leucocephala***

☆ Non leguminous nitrogen fixting trees — ***Alnus nepalensis***

☆ Fast growing forest tree — ***Eucalyptus* species**

☆ Bio-drainage plants — ***Eucalyptus tereticornis***

☆ Fodder producing tree — ***Prosopis cineraria***

☆ Green manuring tree — ***Thespesia populnea***

☆ Tree species suitable for alley cropping — ***Cassia siamea, Leucaena* and *Sesbania***

☆ Fuel wood tree — ***Albizia lebbeck***

☆ Multipurpose tree species — ***Albizia lebbeck***

☆ Miracle forest tree — **Subabul**

☆ Subabul provides — **Fuel, Fodder, Pulpwood, Timber**

☆ Ratanjot and Karanj are — **Biofuel plants**

☆ Oil percentage in Ratanjot (*Jatropha* spp.) — **35 per cent**

☆ Process in which the branch of a plant is cut off in order to produce a flush of new shoots — **Pollarding**

☆ Pollarding is done at — **2 m height from ground**

☆ Bamboo is propagated by — **Seedling, Suckers and Clumps**

☆ Total life span of bamboo tree — **30 years**

☆ Study of forests and woods is termed as — **Silvology**

☆ Match stick industry is based on — **Poplar tree**

☆ Which tree is known as Flame of forest — *Butea monosperma*

☆ Xylose is known as — **Wood sugar**

☆ Ozone layer in atmosphere absorbs — **Ultra violet rays**

☆ Kyoto protocol is related to — **Green house gas minimization**

☆ Carbon dating technique is used for determination of age of — **Fossils**

☆ Arboretum is a — **Botanical garden with trees and shrubs**

☆ Main stem of a tree is called — **Bol**

☆ Sunder lal bahuguna is related to — **Chipko movement (1973)**

☆ Chipko movement originated in — **Tehri Garwal of Uttarakhand**

☆ Which state having highest forest area in India — **Madhya Pradesh**

☆ Which forest type found maximum in India — **Tropical dry deciduous forest**

☆ Study of interactions between living organism and environment is called — **Ecology**

☆ Sum total condition in which organisms live is called as — **Environment**

☆ Large portions of the earth with similar climate, soil, plant and animal life community is known as — **Biosphere**

☆ Optimum area under forest required — **33 per cent of total geographical area**

☆ Most important agroforestry practice is — *Acacia leucophloea* + *Cenchrus setigerus*

☆ Oldest agro forestry practice known as — **Shifting cultivation**

☆ Shifting cultivation causes — **Deforestation**

☆ Taungya system means — **Hill cultivation**

☆ Which test has self purification capacity of water body — **Biochemical Oxygen Demand (BOD) test**

☆ Belt of trees and or shrubs maintained for the purpose of shelter from wind, sun, snow, drift — **Shelterbelts**

☆ Protective plantation in a certain area, against strong winds, it is usually comprised of a few rows of trees — **Wind breaks**

☆ Botanical survey of India (BSI) established in — **1890**

☆ Zoological survey of India (ZSI) established in — 1916

☆ Head quarter of ZSI — **Kolkata**

☆ Tiger project launched in — **1 April 1973**

☆ Environmental information system (ENVIS) set up in — 1982

☆ National museum of natural history (NMNH) setup in — 1978

☆ National museum of natural history (NMNH) located in — **New Delhi**

☆ Ganga action plan phase-I initiated in — 1985

☆ Ganga action plan Phase-II merged with — **National Reserves Conservation Plan (NRCP)**

☆ Study of trees and shrubs is called — **Arboriculture**

☆ Study of changes of ice and all shapes is called — **Glaciology**

☆ Study of environmental ecology of individual species — **Auteocology**

6
Soil Science

☆ Father of soil science — **Vasily Vasili'evich Dokuchaev (Geologist)**

☆ Father of agriculture chemistry — **J. von Liebig**

☆ Law of minimum was given by — **J. von Liebig (1840)**

☆ Crop logging technique was given by — **H. F. Clements**

☆ A value was proposed by — **Fried and Dean (1952)**

☆ NBSS and LUP centre situated at — **Nagpur**

☆ Dokuchalev gave the factor of — **Soil formation**

☆ Soil word is derived from — **Latin**

☆ Study of origin, classification, morphology of soil is known as — **Pedology**

☆ Natural soil aggregates/mass are known as — **Peds**

☆ Study of soils in relation to crop growth is — **Edaphology**

☆ Study of rock is known as — **Petrology**

☆ Soil is formed from — **Weathering of rocks**

☆ Disintegration + Decomposition is called — **Weathering**

☆ Available soil nutrient, determined in terms of a standard fertilizer used is called — **A value**

☆ Substance added to soils for the improvement of their condition known as — **Amendments**

☆ Concentration of water in soil — **25 per cent**

★ Concentration of air in soil — **25 per cent**

★ Concentration of organic matter in soil — **5 per cent**

★ Concentration of mineral in soil — **45 per cent**

★ The soils having at least 20 per cent organic matter are known as — **Organic soils**

★ Mechanical analysis of soils separation is done by — **Hydrometric method**

★ A vertical section of soil through all its horizons is — **Soil profile**

★ Physical property which can not be changed — **Soil texture**

★ How many group in soil texture — **12 (Sand, Loamy sand, Sandy loam, Loam, Silt, Sandy clay loam, Clay loam, Silty clay loam, Sandy clay, Silty clay, Clay)**

★ Arrangement of soil particles and their aggregate into certain defined pattern is — **Soil structure**

★ Types of soil structure are — **Platy, Columnar, Angular, Spheroidal**

★ The capacity of the soil to change its shape under moist conditions — **Soil Plasticity**

★ Process of deposition of soil material in the lower layers is called — **Illuviation**

★ Movement and removal of material in solution from entire solum is called — **Eluvialtion**

★ Horizontal layers in a soil profile are called — **Horizons**

★ Typically Horizon are 4 types — **O, A, B, C**

★ A+B horizons are combindlly called as — **Solum**

★ A+B+C horizons combindlly called as — **Regolith**

★ Development of all horizons in soil is called — **Horizonation**

★ Which horizon is absent in arable land — **"O" horizon**

★ Top most mineral horizon is — **"A" horizon**

★ Which horizon is called fertile zone — **"A" horizon**

★ illuvial horizone is — **"A$_2$" horizon**

★ Alluvial horizone is — **"B" horizon**

★ C horizon consists of — **Unconsolidated parent materials**

☆ Below C horizon is found R layer which
is known as — **Bed rock**

☆ Maximum Eluviation horizone is — **"E" horizone**

☆ One layer wherein soil materials are removed whereas
illuvial layer is one wherein soil materials removed from
other layers are deposited is called — **Elluvial layer**

☆ Number of soil order are — **12**

☆ According to USDA, soils have been — **10 orders**
divided into **(Entisol, Vertisol, Inceptisol,**
Acidisol, Mollisol, Spodosol,
Alfisol, Ultisol, Oxisol, Histosol)

☆ Latestly added 11th order is Andosol and
12th order is — **Gelisol**

☆ Andosol is found in — **Volcanic cruption area**

☆ Gelisol is found in — **Arctic regions**

☆ Two largest order in India are — **Inceptisol followed by Entisol**

☆ Red soil also known as — **Alfisol**

☆ Red colour in red soil is due to presence of — **Various oxide of**
iron or Ferric oxides

☆ Lime and low soluble salts are absent in which soil — **Alfisol**

☆ Kaolinite (1:1 type clay minerals) rich in which soil — **Alfisol**

☆ Red soil formed from — **Ancient crystalline and**
metamorphic rocks

☆ Parent material for red soil is mostly — **Granite**

☆ pH of alfisols varies from — **6.0 to 7.5**

☆ Night soil also known as — **Poudrette**

☆ Black soil also known as — **Vertisol**

☆ Base exchange capacity of deep black soil is — **Quite high**

☆ Montmorillonite type clay minerals rich in which soil — **Vertisol**

☆ Alluvial soil also known as — **Entisol**

☆ pH of entisol varies from — **7.0 to 8.0**

☆ In alluvial soil P and K found in sufficient
amount but deficient in — **N and organic matter**

☆ Laterite soil also known as — **Ultisol**

☆ pH of ultisol varies from — **5.0 to 6.0**

☆ Kaolinite type clay minerals rich in which soil — **Ultisol**

☆ In laterite soil deficient in — *P, K, Ca, Zn, B etc*

☆ Shifting cultivation is mostly practised in which soil area — **Laterite soil**

☆ Histosol is also known as — **Organic soil**

☆ Kankar nodules are found in mostly in — **Red soils**

☆ Young soil also known as — **Inceptisol**

☆ Recently formed soil order is — **Entisols**

☆ Most dominant soil order of India — **Entisols**

☆ Black soil belongs to the soil order — **Vertisol**

☆ Soil having more than 30 per cent organic matter is placed in — **Histosol**

☆ Most important soil group of India — **Alluvial soil**

☆ Soil deficient in nitrogen content — **Black soil**

☆ Newly formed alluvial soil is — **Khadar**

☆ Black soil shows black colour due to compound of — *Mn*

☆ Vertical cracks are major problem in — **Deep black soil**

☆ Total pore space highest soil is — **Clay soils**

☆ Rocks are divided into 3 type — **Igneous rocks, Sedimentary rocks, Metamorphic rocks**

☆ Granite, Syenite and Basalt are type of rocks — **Igneous rocks**

☆ Dolomite, Lime stone, Sand stone are type of rocks — **Sedimentary rocks**

☆ Metamorphic rocks are formed from — **Igneous rocks and Sedimentary rocks**

☆ Order of occurrence — **Feldspars>Quartz>Mica>Limestone**

☆ Quartzite, Marble, Gneiss, Graphite, Schist and Slate are type of rocks — **Metamorphic rocks**

☆ Quartzite is formed from — **Quartz or Sandstone**

☆ Marble is formed from — **Limestone**

☆ Gneiss is formed from — **Granite**

☆ Graphite is formed from — **Coal**

☆ Slate is formed from — **Shale**

☆ Quartz, Mica and Feldspar are — **Primary minerals**

☆ Major source of magnesium (Mg) is — **Dolomite**

☆ Main source of molybdenum (*Mo*) is — **Olivine**

☆ Main source of Manganese (*Mn*) is — **Pyrolusite**

☆ Main source of phosphorus (*P*) is — **Appetite**

☆ Main source of potassium (*K*) is — **Orthoclase**

☆ Main source of boron (*B*) is — **Tourmaline**

☆ Chief constituent of sandy fraction — **Quartz**

☆ Most resistant to weathering — **Quartz**

☆ Least resistant to weathering — **Calcite**

☆ Weathering mineral, having most stable soil structure — **Kaolinite**

☆ Which mineral is a source of phosphorus and boron in soils — **Apetite**

☆ Hydroxide act as cementing agent in binding the soil particles together — *Fe* **and** *Al*

☆ Kaolinite, Halloysite and Dickite are type of minerals — **1:1 type mineral**

☆ Montmorillonite, Vermiculture and Illite are type of minerals — **2:1 type mineral**

☆ Chlorite is type of mineral — **2:1:1 type minerals**

☆ Mica is type of mineral — **Non Expanding type mineral**

☆ Alfisols, Ultisols, Oxisols are under the group of — **Laterite soil**

☆ Rocks gets broken in pieces due to temperature is called — **Exfoliation**

☆ Most dominant mineral on earth crust — **Feldspar (48 per cent)**

☆ Black soil of central high land and plateaus have high content of — **Montmorillonite**

☆ Cation exchange capacity (CEC) is highest in — **Montmorillonite**

☆ Cation exchange capacity (CEC) of Montmorillonite is — **100 me/100 g**

☆ C:N ratio of normal soil is — **12:1**

☆ C:N ratio of F.Y.M. is — **100:1**

☆ C:N ratio of humus/organic matter is — **10:1**

☆ C.E.C of humus is — **150-300 C mol(P⁺)/kg soil**

☆ C:N ration of micro organism is — **4:1 to 9:1**

☆ C:N ratio of legumes is — **20:1 to 30:1**

☆ C:N ratio of cereals is — **90:1**

☆ C.E.C of vermiculture is — **80-150 C mol(P⁺)/kg soil**

- ☆ C.E.C. of montmorillonite is — **80-100 C mol(P⁺)/kg soil**

- ☆ Formula of porosity of soil — **= 100 – (Bulk density/ Particle density) × 100**

- ☆ Formula of bulk density — **Weight of dry soil/ Volume of soil(solids + pores)**

- ☆ Formula of particle density — **Weight of dry soil/ Volume of soil solid**

- ☆ Particle density of soil is also known as — **True density**

- ☆ Particle density of soil — **2.65 g/cm³**

- ☆ Mass per unit volume is called — **Bulk density**

- ☆ Bulk density of soil — **1.33 g/cm³**

- ☆ Diameter of sand soil is — **0.2 – 2 mm**

- ☆ Diameter of silt soil is — **0.002 - 0.02 mm**

- ☆ Diameter of clay soil is — **below 0.002 mm**

- ☆ Porosity percentage of sandy soil is — **30 per cent**

- ☆ Porosity percentage of clay soil is — **50-60 per cent**

- ☆ Porosity percentage of loamy soil is — **40-50 per cent**

- ☆ Irrigation efficiency of loamy soils — **70 per cent**

- ☆ Particle more than 250 mm in diameter is known as — **Stone**

- ☆ Which soil is most suitable for most of crops — **Sandy loam soil**

- ☆ Which soil structure is best suitable for cultivation is — **Crumby structure**

- ☆ Soil structure proving less porosity in soil — **Platy**

- ☆ Soil colour is determined by — **Munsell colour chart**

- ☆ Humus contains — **C- 50 per cent, H-5 per cent, O-35 per cent, N-5 per cent**

- ☆ Weight of soil furrow slice/surface soil (0-15 cm) — **2×10⁶ kg/ha**

- ☆ The soil having PD 5.00 g/cc and BD 2.50 g/cc will have porosity per cent — **50**

- ☆ What is the moisture per cent on dry weight basis if field soil sample weighing 60 g, lost 12 g on over drying — **25 per cent**

- ☆ Loss of N_2 in the form of NH_3 in alkaline medium is known as — **Volatilization**

- ☆ Zonal soil farming process are — **Calcination, Podzolization, Laterisation**

- ☆ Podzolization occurs in — **Cold humid temperate conditions**

☆ Process of moving out of sesquioxide is known as — **Podzolization**

☆ Process of soil formation in which sesquioxides are leached from A horizon and silica is left out in upper layer is called — **Podzolization**

☆ Laterisation occurs in — **Warm humid tropical conditions**

☆ Process of soil formation in which silica is remove from A horizon and sesquioxides are left out in A horizon is called — **Laterisation**

☆ Process of mixing of soils is known as — **Pedoturbation**

☆ Accumulation of soluble salts in soil is called — **Salination**

☆ Unconsolidated product of rocks by weathering is called — **Regolith**

☆ Deflocculation of soil particles occurs by — **Na**

☆ Wood is mainly decomposed by — **Actinomycetes**

☆ Smell of soil after fresh shower is due to — **Actinomycetes**

☆ Attraction of solid surface for water molecules is called as — **Adhesion**

☆ Density of soil water is maximum at — **4^0C**

☆ Surface tension of soil water is at 25^0C — **72.7 dyne/cm^2**

☆ Mechanical analysis of soil is estimated by — **Stock`s law**

☆ pH value varies from — **0-14**

☆ Infiltration rate is relatively higher in — **Sandy soil**

☆ A value is used for the assessment of available in soil is — **P and S**

☆ Force of attraction that binds the molecules of different kinds is called — **Adhesion**

☆ Map scale used in reconnaissance soil survey — **1:15000**

☆ Which plant nutrient is leached maximum — **Ca**

☆ Excess Na causes — **Ca deficiency**

☆ More Ca is present in — **Older leave**

☆ Excess Mn causes — **Fe deficiency**

☆ Order of occurrence — **Bacteria > Actinomycetes > Fungi > Algae**

☆ Order of occurrence — **Feldspars > Quartz > Mica > Lime stone > Hornblende and augite > Olivine and serpentine**

☆ Excellent water holding capacity is — **Loam soil**

7
Fertilizers

☆ Functional or Metabolic nutrients was proposed by — **Nicholas (1961)**

☆ Criteria of essentiality was proposed by — **Arnon and Stout (1939)**

☆ Method used for the determination of lime requirement of an acids soil is — **Shoemaker`s method**

☆ Nutrient mobility concept was propounded by — **Bray**

☆ Essentially of *N* was established by — **De Saussure**

☆ Arnon and Stout discovered the essentially of — *Mo*

☆ Total number of essential nutrients for plants — 17

☆ Recently added nutrient is — Ni

☆ Total number of functional nutrients — **21 (essential-17 + Co,Si,Na,V)**

☆ Highly mobile nutrients are — **N, P, K**

☆ Macro nutrients are — **N, P, K, Ca, Mg, S, C, H, O (>100 µg/g dry matter)**

☆ Micro nutrients are — **Zn, B, Fe, Cu, Mo, Mn, Cl (100 µg/g dry matter)**

☆ Primary nutrienrs are — **N, P, K**

☆ Secondary nutrients are — **Ca, Mg, S**

☆ Mobile elements are — **N, P, K, Zn, Na, Mg, Mn, Mo, Cl**

☆ Immobile nutrients in plants are — **B, Ca, Cu, Fe, S**

☆ Immobile nutrients in soil is — **P**

☆ Metal nutrient are — **Zn, Mg, Mn, Ca, Cu, K, Fe**

☆ Moderate mobile nutrient is — **Zn**

☆ Elements provide basic structure — **C, H, O**

☆ Beneficial elements are — **Co, Na, Si, Ni, Ru, Sr, As, Ru, Sl**

☆ Ultra micro nutrient is — **Mo**

☆ Cataion nutrients are — **N, K, Fe, Mn, Cu, Zn**

☆ Anion nutrients are — **P, B, Mo, H_2PO_4**

☆ Cobalt is essential element for — **Legumes**

☆ Silicon and Nickel is essential element for — **Paddy, Maize**

☆ Energy exchange elements are — **Hydrogen and Oxygen**

☆ Micronutrient is also known as — **Trace elements, Oligo elements, Spurne elements**

☆ Acid rain water contains — **NO_2^-, SO_4^-, H^+**

☆ pH of acid rain water — **5-7**

☆ Plant take thenutrient in the form of — **Ions**

☆ Potassium and Sodium is determined by — **Flame Photometer**

☆ Available potassium is determined by — **Flame photometer in soil extract**

☆ Sodium is estimated by flame photometer in a — **Blue flame**

☆ Luxury consumption or maximum uptake nutrient is — **Potassium**

☆ Element contributing to the disease and drought resistant — **Potassium**

☆ Ni and Co are essential for — **Legumes crop**

☆ Si is essential for — **Paddy and Maize**

☆ Structural component of Vitamin B_{12} is — **Co (Cobalt)**

☆ Which element is the constituent of chlorophyll — **Mg (Magnesium)**

☆ Which element is associated to aromatic compounds — **S (Sulphur)**

☆ S (Sulphur) nutrient is essential for — **Oil seed crops**

☆ Excess vegetative growth is due to — **High supply of N_2**

☆ Na is essential for — **Sugar beet**

☆ Sulphur is an essential constituent of the sulphur containing amino acids, these are — **Methionine, Cystein**

☆ When super phosphate is used during compost making, is called — **Super Compost**

☆ Equivalent acidity of $Ca(NO_3)_2$ is — **21**

☆ Equivalent acidity of N_ANO_3 is **–29**

☆ Equivalent acidity of Ammonia Nitrate is — **60**

☆ Equivalent acidity of DAP is — **77**

☆ Equivalent acidity of Urea is — **80-85**

☆ Equivalent acidity of ammonium phosphate is — **96**

☆ Equivalent acidity of $(NH_4)_2SO_4$ is — **110**

☆ Equivalent acidity of NH_4Cl — **128**

☆ Equivalent acidity of Anhydrous Ammonia is — **148**

☆ Urea is a which type of fertilizer — **Organic fertilizer**

☆ Cheapest N_2 contain fertilizer suitable for foliar sparay — **Urea**

☆ Cheapest N_2 contain fertilizer is — **Urea**

☆ Amide form of N_2 fertilizer is — **Urea**

☆ First product of urea hydrolysis — **Ammonium carbamate**

☆ Maximum allowable biurate content of Urea is — **1.5 per cent**

☆ Least hygroscopic fertilizer is — **DAP**

☆ Major P_2O_5 fertilizer in India is — **DAP**

☆ Which fertilizer is most beneficial for alkali soils — **Ammonium Sulphate**

☆ Chemical formula of urea — **$Co(NH_2)_2$**

☆ Chemical formula of Ammonium sulphate — **$[(NH_4)_2.SO_4]$**

☆ Chemical formula of Potassium Nitrate — **KNO_3**

☆ Chemical formula of Single super phosphate — **$Ca(H_2PO_4)$**

☆ Concentration of foliar spray of urea varies between — **1-2 per cent**

☆ Nitrate fertilizer should be used in — **Oxidized zone**

☆ Ammonical fertilizer should be used in — **Reduced zone**

☆ All nitrogenous fertilizer are acidic in nature except — **CAN, NO_3, CaCN**

☆ Element available in both anion and cation form is — **Nitrogen**

☆ Which crop used Nitrogen as a Ammonical form — **Paddy and Potato**

☆ Maximum amount of potassium fertilizer is applied in — **Potato**

☆ Nutrient required for quality maintenance in potato — **Potassium**

☆ Which form of N_2 is preferable for saline soils — **Nitrate form**

☆ Phosphorus is taken by plant in the form of — **Phosphoric acid**

☆ Phosphorus fixation is most probable in — **Laterite soil**

☆ Phosphorus is available at — **6.0 - 7.0 pH**

☆ Phosphorus is extracted by — **Olsen's method and Bray no.1 method**

☆ Element involve in energy transfer and storage in plant — **Phosphorus**

☆ Organic carbon is determined by — **Morgan's method and Walkey and black method**

☆ Source of N for plants for absorption — **NO_3**

☆ Source of P for plants for absorption — **HPO_4^{-2}, $H_2PO_4^{-}$**

☆ Source of B for plants for absorption — **$B_4O_7^{-2}$, $H_2BO_3^{-}$**

☆ Source of Mo for plants for absorption — **MoO_4^{-2}**

☆ Thiourea, Oxamide, Dicyandiamide, Neem cake are — **Nitrogen inhibitor**

☆ Water soluble phosphatic fertilizer are — **DAP, SSP, DSP, TSP, Mono ammonium Phosphate etc.**

☆ Citrate soluble phosphatic fertilizer are — **Basic slag, Dicalcium phosphate etc.**

☆ Water and Citrate Insoluble phosphatic fertilizer are — **Rock phosphate, Raw bone meal etc.**

☆ Gromor, Neem Coated Urea, Urea Formaldehyde are — **Slow released fertilizer**

☆ Calcium Ammonium Nitrate (CAN) is also known as — **Kisan Khad**

☆ Neutral fertilizer also called — **Kisan Khad**

☆ Humic acid is soluble in — **Alkali solution**

☆ Osmotic regulation is maintained by — **K^{+}**

☆ Acid tolerant crop is — **Paddy**

☆ High lime requirement crops are — **Soybean and Sugar beet**

☆ Highly salt tolerant crops are — **Barley and Sugar beet**

☆ Rock phosphate is used in — **Acidic soil**

☆ Which crop prefer for acidic soil — **Paddy, Patato, Tea**

☆ Fertilizer having less than 25 per cent of the primary nutrients known as — **Low analysis fertilizer**

☆ Micro nutrient deficient in Indian soils is — **Zinc**

☆ Major nutrient deficient in Indian soils is — **Nitrogen**

☆ Application of clay to sandy soils is known as — **Marling**

☆ Phosphatic fertilizer suitable for acid soil — **Bone meal**

☆ Azolla is a — **Aquatic fern**

☆ Nitrogen fixation in paddy field is carried out by which BGA — **Azolla**

☆ What is amount of N fixed by Azolla — **25-30 kg/ha**

☆ What is amount of N fixed by BGA — **15-45 kg/ha**

☆ PSB, Rhizobium, Azotobactor, Azospirillum, BGA are — **Biofertilizer**

☆ One packet of Rhizobium culture contains — **150 g**

☆ One packet of PSB culture contains — **250 g**

☆ One packet of Azotobactor culture contains — **150 g**

☆ Azotobactor is used for — **Paddy, Wheat, Sugarcane, Cotton**

☆ Azotobactor is — **Aerobic bacteria**

☆ Azatobactor fixes atmospheric nitrogen — **20-30 kg/ha**

☆ Azotobactor and Azospirillum fixes atmospheric nitrogen — **Non-symbiotically or Asymbiotic**

☆ Free living nitrogen fixing organism is — **Azatobactor**

☆ Rhizobium is used for — **Pulse crops**

☆ Essential element required by the N fixing bacterium Rhizobium — **Mo**

☆ Rhizobium fixes atmospheric nitrogen — **50-100 kg/ha**

☆ *Rhizobium leguminosarum* is used for — **Pea, Lentil, Lathyrus**

☆ *Rhizobium Japonicum* is used for — **Soybean, Groundnut, Cow pea**

☆ Rhizobium phaseoli is used for — **Beans**

☆ Rhizobium melilots is used for — **Alfalfa, Fenugreek**

☆ *Rhizobium trifoli* is used for — **Clover/Berseem**

☆ PSB is used for — **All crops**

☆ Clostridium is — **Anaerobic bacteria**

- ☆ Azospirillum is used for — **Sorghum**
- ☆ Phosphate solubilizing fungi is — **Aspergillus, Penicillium**
- ☆ Solubility of rock phosphate can be improved by — ***Bacillus polymexa***
- ☆ VAM belongs to the group of — **Fungi**
- ☆ Optimum temperature and pH for nitrifying bacteria is — **Temp.= 30-35^{0}C and pH = 6.5 - 7.5**
- ☆ Most out standing green manure crop — **Sun hemp**
- ☆ Fastest N_2 fixing plant — ***Sesbania rostrata***
- ☆ Low status of N_2 in soil when it is less than — **250 kg/ha**
- ☆ Analyzing process for determination of available N_2 — **Alkaline permanganate method**
- ☆ KCl is denoted to — **Murate of Potash**
- ☆ Boric Acid contain — **17 per cent B**
- ☆ Copper Sulphate ($CuSO_4.5H_2O$) contain — **24 per cent Cu**
- ☆ Cuprous Oxide (Cu_2O) contain — **89 per cent Cu**
- ☆ Which fertilizer highest N_2 per cent — **Anhydrous Ammonia (82 per cent)**
- ☆ Urea Formaldehyde contain — **38-42 per cent N_2**
- ☆ Thio urea contain — **36.8 per cent N_2**
- ☆ Manganous Oxide contain — **41-68 per cent Mn**
- ☆ Zinc Dust contain — **60-80 per cent Zn**
- ☆ Gypsum contains — **18.6 per cent S and 29.2 per cent Ca**
- ☆ Ammonium polyphosphate contains — **15 per cent N and 62 per cent P**
- ☆ explosive fertilizer is — **Ammonium Nitrate**
- ☆ Peaty soils are generally deficient of — **Cu**
- ☆ Marshy soils are generally deficient of — **Zn**
- ☆ Process of decomposition of organic matter is termed as — **Humification**
- ☆ Mass of rotted organic matter made from waste — **Compost**
- ☆ Organic matter rich compost made by use of earth worms is called — **Vermicompost**
- ☆ A practice of turning undecomposed fresh green plant tissue into the soil to improve fertility status and physical structure of the soil is called — **Green manuring**

☆ Green manure crops are turned in the field at the stage of — **flowering**

☆ Green manure crops contributes N_2 — **50-175 kg/ha**

☆ Most widely used green manure crop — **Sun hemp (*Crotalaria juncea*)**

☆ Green manure crop having both stem and root nodulation — *Sesbania rostrata*

☆ Green leaf manuring crop — **Karanj and Ipomea**

☆ Crop oil cake, which has highest nitrification rate — **Groundnut**

☆ Potassic fertilizer suitable for fertigation — **Potassium Nitrate**

☆ Application of fertilizer along with irrigation water — **Fertigation**

☆ Which nutrient can be applied by fertigation — **Nitrogen and Sulphur**

☆ Conversation of Ammonia (NH_3^+) to Nitrite (NO_2^-) in the soil by — **Nitrosomonas**

☆ Conversation of NO_2^- (Nitrite) to NO_3^- (Nitrate) by — **Nitrobactor**

☆ Conversation of soil nitrate into gaseous nitrogen is — **Denitrification**

☆ Water logged paddy field atmospheric nitrogen can be fixed to the soil by — **BGA**

☆ Phosphate solubilizer species of micro-oraganism is — **Pseudomonas**

☆ Mychorriyza increase availability of — **Phosphorus**

☆ Zinc solubilizing bacterial biofertilizer — **Azozink**

☆ Which nutrient maximum uptake by the plant — **K⁺**

☆ Situation in which a crop needs more of a given nutrient yet has shows no deficiency symptoms — **Hidden hunger**

☆ Residual effect of urea on soil reaction is — **Acidic**

☆ Fertilizer that released an acidic effect in the soil is called — **Acid forming fertilizer**

☆ Urea, Ammonium sulphate, Ammonium chloride, Anhydrous ammonia are — **Acid forming fertilizer**

☆ Metallic cation mostly lack in — **Acid forming fertilizer**

☆ Fertilizer not produced in India — **MOP (Muriate of Potash)**

☆ Maximum saline and alkali soil are found in which state — **U.P.**

☆ Alkali soils are generally found in which climate — **Arid and Semi arid climate**

☆ Smell of soil after fresh showers is due to — **Actinomycetes**

☆ Order of maximum in earth — **Bacteria>Actinomycetes>Fungi>Algae**

☆ 1 tone of cattle dung give — N=2.95 kg, P=1.59 kg, K=2.95 kg

☆ Available nitrogen is determined by — **Alkaline permanganate method**

☆ Cereals straw is a particularly rich source of — **Potash**

☆ Central Fertilizer Quality Control Institute situated at — **Faridabad (Haryana)**

☆ Ammonium oxalate is also called — **Grigg`s reagent**

☆ V shape pattern of yellowing shows — **N_2 deficiency**

☆ Deficiency symptoms of N, P, K, Mg and Mo appear in — **Older leaves**

☆ Deficiency symptoms of Fe, Mn, Cu, and S appear in — **New leaves**

☆ Deficiency symptoms of Zn appear in — **Old and New leaves**

☆ Deficiency symptoms of Ca and B appear in — **Terminal buds**

☆ Excess of N, P and K causes deficiency of — **Cu**

☆ Excess of Ca causes deficiency of — **P**

☆ Resetting and excess gumming due to — **Cu deficiency**

☆ Deficiency symptoms of nitrogen occurs first on — **Lower leaves**

☆ Cereal crops "V" shaped yellowing at the tip of the lower leaves shows — **N deficiency**

☆ Deficiency appears as short internodes in plant — **N**

☆ Purple coloration appeared in leaves due to deficiency of — **P**

☆ Scorching and burning on marginals of bottom leaves and irregular fruit development of plant are most common symptom of — **K deficiency**

☆ Deficiency symptoms of calcium occurs first on — **Terminal leaves**

☆ Interveinal chlorosis occurs due to — **Fe and Mg deficiency**

☆ Complete interveinal chlorosis due to — **Mn deficiency**

☆ Brittle leaf due to — **Deficiency of Ca**

☆ Chlorosis in between the veins and veins remain green shows — **Mg deficiency**

☆ Tip burn, margin scorching shows — **K deficiency**

☆ Tip burn of paddy is due to — **O_2 and N_2 deficiency under submerged condition**

☆ Upper leaves will show chlorosis on mid rib, veins green and dead spot occur in all parts of leaf show — **Deficiency of Zn**

☆ Zn toxicity is reduced by addition of — **Super phosphate**

- ☆ White bud and white tip of maize is due to — **Deficiency of Zn**
- ☆ Khaira disease of paddy is due to — **Deficiency of Zn**
- ☆ Reclamation disease is due to defiency of — **Deficiency of Cu (Copper)**
- ☆ Pahala blight of sugarcane is due to — **Deficiency of Mn (Manganese)**
- ☆ Marsh spot of pea is due to — **Deficiency of Mn (Manganese)**
- ☆ Grey speck of oat is due to — **Deficiency of Mn (Manganese)**
- ☆ Interveinal yellowing of younger leaves — **Deficiency of Mn (Manganese)**
- ☆ Grey spot on leaves is due to — **Deficiency of Mn (Manganese)**
- ☆ Top sickness of tobacco is due to — **Deficiency of B (Boron)**
- ☆ Heart rot or Brown heart of turnip and sugar beet is due to — **Deficiency of B (Boron)**
- ☆ Sickle leaf disease due to — **Deficiency of P**
- ☆ Tea yellow disease due to — **Deficiency of S**
- ☆ Downward cupping of leaves in Tobacco and Tea shows — **Deficiency of S**
- ☆ Failure of terminal bud and root tip due to — **Deficiency of Ca**
- ☆ Whip like structure appeared in terminal bud — **Deficiency of B**
- ☆ Burning quality of tobacco decreased due to — **Deficiency of Cloride**
- ☆ Translucent spots of irregular shape between veins shows — **Deficiency of Mo**
- ☆ Boron is harmful for plants having concentration — **More than 3 ppm**

8

Biochemistry

- ☆ Biochemistry word derived from — **Greek word**
- ☆ Father of agricultural biochemistry — **Justus von Liebig**
- ☆ First used term biochemistry — **Neuberg (1903)**
- ☆ pH concept introduced/discovered by — **Sorensen (1909)**
- ☆ Protein term derived from — **Greek word**
- ☆ Protein term was coined by — **Moulder (1840)**
- ☆ Protein is discovered by — **Berzelius (1938)**
- ☆ Enzyme term was given by — **W. Kuhne (1867)**
- ☆ Enzyme is discovered by — **Buckner (1897)**
- ☆ Nucleic acids first discovered by — **Friedrich Meischer (1868) in Puss cell**
- ☆ Base composition of DNA discovered by — **Chargaff (1953)**
- ☆ Double helix model of DNA was proposed by — **Watson and Crick (1953)**
- ☆ Vitamin was discovered by — **Funk (1911)**
- ☆ Plant grown in salt water is known as — **Halophyte**
- ☆ Saline soil also called — **White alkali and Solan chalk**
- ☆ Alkaline soil also called — **Black alkali and Solanetz**
- ☆ Saline – alkali soil also called — **Usar**

☆ What is used for reclamation of saline soils — **Pyrite**

☆ Leaching is used for treatment of — **Saline soils**

☆ What is used for reclamation of Acidic soils — **Lime stone (CaCO$_3$)**

☆ What is used for reclamation of alkaline/sodic soils — **Gypsum (CaSO$_4$.2H$_2$O)**

☆ What is used for reclamation of saline soils — **Pyrite (FeS$_2$)**

☆ Soil PH >8.5 indicates soil is — **Alkaline**

☆ When Ec >4 (ds/m), ESP <15 per cent, pH <8.5, soil is called as — **Saline soil**

☆ When Ec <4 (ds/m), ESP >15 per cent, pH >8.5, soil is called as — **Alkali soil**

☆ When Ec >4 (ds/m), ESP >15 per cent, pH >8.5, soil is called as — **Saline-alkaline soil/Sodic**

☆ Chemical formula of Gypsum is — **CaSO$_4$.2H$_2$O**

☆ Chemical formula of Pyrite is — **FeS$_2$**

☆ Exchangeable sodium percentage (ESP) is also known as — **Soluble sodium percentage (SSP)**

☆ Sodium Adsorption Ratio (SAR) is equal to — $= Na^+/\sqrt{(Ca^{+2}+Mg^{+2})/2}$

☆ In acidic soils, the toxicity increased due to — **Increasing of Aluminium**

☆ Soils which have pH <4 are known as — **Cat soil**

☆ Soils which have organic matter and Na are known as — **Black alkali soils**

☆ Highest area of acidic soil in India — **West Bengal**

☆ Potato scab disease favours — **Alkaline soils**

☆ Fertilizer which destroys soil aggregates — **Sodium nitrate**

☆ Which of the plant species can be suggested on saline soil — *Haloxylon salicornium*

☆ Polymer of amino acid — **Proteins and Enzyme**

☆ Protein that contains only amino acid — **Simple protein**

☆ Regulatory protein are — **Insulin**

☆ Transport protein are — **Haemoglobin and Myoglobin**

☆ Structural proteins are — **Collagen and Elastin**

☆ Most abundant protein present in the world — **Rubisco**

☆ Muscle protein is known as — **Collagen**

☆ Silk protein is known as — **Fibrolin**

☆ Wheat protein is known as — **Gluten**

☆ Soybean protein is known as — **Glycinin**

☆ Rice protein is known as — **Oryzein**

☆ Maize protein is known as — **Zein**

☆ 1st enzyme which was discovered by yeast — **Zymase**

☆ Chloropicrin known as — **Tear gas**

☆ Enzyme involve in biological nitrogen fixation — **Nitrogenase**

☆ Component of nitrogen reduction enzyme is — **Nitrogenase**

☆ Component of nitrate reduction is — **Molybdenum**

☆ Apoenzyme + Prosthetic group — **Holoenzyme**

☆ Enzyme without prosthetic group — **Apo-enzyme**

☆ Enzyme which exist in multiple forms within single specing of an organism — **Isoenzyme**

☆ Non protein component of the enzyme — **Co-enzyme**

☆ Total number of essential amino acids are — **10**

☆ Full form of DNA — **Deoxyribose Nucleic Acid**

☆ Deoxyribose sugar + Nitrogenous base is — **Nucleoside**

☆ Deoxyribose sugar + Nitrogenous base + Phosphate group — **Nucleotide**

☆ Single stranded DNA found in — **Bacteriophage**

☆ Form of DNA present in living organisms — **B form**

☆ RNA that transfer amino acids form cytoplasm to ribosome — **mRNA**

☆ Most abundant form of RNA that constitute 80 per cent parts of cellular — **tRNA**

☆ Which nucleic acid is necessary for protein biosynthesis — **RNA**

☆ Nitrogen base of DNA — **Adenine, Thymine, Guanine, Cytosine**

☆ Nitrogen base of RNA — **Adenine, Thymine, Guanine, Uracil**

☆ Total well define vitamins are — **13**

☆ Vitamin B complex (B_1, B_2, B_3, B_6, B_{12}) and Vitamin C are — **Water soluble vitamins**

☆ Vitamin A,D,E,K are — **Fat soluble vitamins**

☆ Which amino acid deficient in Pulses are — **Methionine**

☆ Which amino acid deficient in Cereals are — **Lysine**

☆ Which vitamin contains metal — **Vitamin B$_{12}$**

☆ reaction of oil/fat with NaOH/KOH as — **Saponification**

☆ number of grams of iodine absorbed by 100 g fat of oil — **Iodine number**

☆ value, used to assess the degree of spoilage (rancidity) of a fat or oil — **Acid number**

☆ Partial substitute for petroleum diesel — **Biodiesel**

☆ Golden rice is rich in — **β carotene**

☆ Water, on freezing, expands per cent by volume — **9 per cent**

☆ Water has maximum density at which temperature — **3.98°C or 39.16°F**

☆ Most abundant bio molecules on earth — **Carbohydrates**

☆ Chemical formula of monosaccharide — **C$_6$H$_{12}$O$_6$**

☆ Carbohydrates can be classified in — **3 categories (Monosaccharide, Oligosaccharide, Polysaccharide)**

☆ Monosaccharide contains — **Glucose, Fructose, Galactose, Ribulose, Xylose, Fructose and Mannose**

☆ Monosaccharides posses reducing property due to the presence of — **Keto group or Free aldehyde**

☆ Monosaccharide that are used as energy source — **Glucose and Fructose**

☆ Glucose is also known as — **Dextrose, Corn sugar**

☆ Biologically active form of glucose — **D-form**

☆ Fructose is also known as — **Fruit sugar**

☆ Sweetest sugar among all the sugars is — **Fructose**

☆ Oligosaccharide contains — **Lactose, Maltose, Sucrose, Cellobiose, Raffinose, Stachyose**

☆ Oligosaccharide classified as — **Disaccharides, Trisaccharides, Tetrasaccharides**

☆ Disaccharides contains — **Lactose, Maltose, Sucrose, Cellobiose**

☆ Trisaccharides contains — **Raffinose**

☆ Tetrasaccharides contain — **Stachyose**

☆ Oligosaccharides that is used in preservation of foods — **Sucrose**

☆ Non reducing type sugar is — **Sucrose**

☆ Reducing type sugars are — **Maltose and Cellobiose**

☆ Glucose and Glactose consisting of — **Lactose**

☆ Sugar presents in milk — **Lactose**

☆ Glycogen present in — **Animal cell**

☆ Polymer of glucose — **Cellulose**

9
Horticulture

☆ Horticulture origin from   **– Latin word**

☆ Study of fruit science is called   **– Pomology**

☆ Pomology origin from   **– Greek word**

☆ Study of vegetable science is called   **– Olericulture**

☆ Olericulture origin from   **– Latin word**

☆ Father of Indian Horticulture   **– M. H. Marigowda**

☆ Father of canning   **– Nicholas Appert**

☆ Study of flower and ornamental plant science is called   **– Floriculture**

☆ Study of fruit and vegetable preservation science is called   **– Postharvest technology**

☆ Method by which food is kept out from spoilage after harvest is called   **– Preservation**

☆ Which operation to controls the shape of plant   **– Training**

☆ Which operation to remove of any excess or unproductive branches, or any other parts of plants   **– Pruning**

☆ Heading back and thinning out are associated with   **– Pruning**

☆ Maturity of sweet cron measured by   **– Succulometer**

☆ More pruning required in   **– Decidious fruit plants**

☆ Which fruit does not required pruning   **– Banana, Papaya, Pineapple**

☆ Most widely used training system for commercial fruits — **Modified Leader System**

☆ Simplest system of fruit planting — **Square System**

☆ Filler tree technology is associated with — **Quincunx System**

☆ Which grafting is used for repairing the plant — **Bridge grafting**

☆ Mango, Banana, Guava, Papaya, Jackfruit, Apple, Sapota are — **Climacteric fruits**

☆ Citrus, Grape, Litchi, Ber, Pineapple are — **Non-Climacteric fruits**

☆ Queen of spices — **Cardamon**

☆ Food of god is — **Cocoa**

☆ Wind breaks always planted in the direction of — **North-West**

☆ Father of systemic pomology — **Decandolle**

☆ APEDA was established in — **1985**

☆ Coffee board was established in — **1942**

☆ ICAR-Indian Institute for Horticulture Research (IIHR) is situated at — **Bangalore in 1968**

☆ National Horticulture Board (NHB) was established at — **Gurgaon (Haryana) in 1984**

☆ National Research Centre on Medicinal and Aromatic Plants is situated at — **Anand (Gujarat)**

☆ CSIR-Central Institute For Medicinal And Aeromatic Plants (CIMAP) situated at — **Lucknow (U.P.)**

☆ ICAR-Central Temperate Horticulture Research Institute (CTHRI) situated at — **Srinagar (J&K)**

☆ International Institute of Horticulture is situated at — **Brazil**

☆ Coconut development board comes under — **Ministry of Commerce**

☆ First All India Coordinated Floriculture Improvement Project was started in — **1971**

☆ Who was the first Deputy Director General of Horticltue, ICAR — **Dr. K. L. Chadha**

☆ Removal of brances and top portion of plant stem along with leaves is called — **Pinching**

☆ Chinese pot layering is ideal for – **Air layering**

I. Cultivation of Fruit Crops

Mango

☆ Botanical name — *Mangifera indica*

☆ Family — **Anacardiaceae**

☆ Origin — **Indo-Burma region**

☆ Also called — **Bathroom fruit, King of fruits, National fruit**

☆ Fruit type — **Drupe**

☆ Edible part — **Mesocarp**

☆ Pollination by — **House fly**

☆ Inflorescence of mango — **Panicle**

☆ Bearing habit — **Terminal**

☆ Mango inflorescence flower contain which type flower — **Male and Hermaphrodite**

☆ Commercially propagated by — **Veneer grafting**

☆ In situ method of propagation — **Soft wood grafting**

☆ Dwarfing root stock in mango — **Totapari red small**

☆ Normally planting space — **10×10 m^2**

☆ Which type of incompatibility is found in mango — **Gametophytic**

☆ Spongy tissue due to — **Convection heats**

☆ Internal fruit necrosis due to — **Boron deficiency**

☆ Mango malformation is caused by — **Fungus (*Fusarium moniliforme*)**

☆ Mango malformation was firstly observed at — **Darbhanga, Bihar (1891)**

☆ Leaf scorch in mango is due to toxicity of — **Chloride**

☆ Deblossoming is done for — **Control of mango malformation**

☆ Paclobutrazol and Kultar chemical used for — **Avoid the alternate bearing**

☆ Fruit drop in mango is controlled by — **Spray of 2,4-D**

☆ How much mango flowers develop fruit and reach maturity — **0.1 per cent or less**

☆ Black tip of mango disorder is due to — **Boron deficiency**

☆ Control of black tip of mango by spray of — **Borax spray**

☆ Amprali variety of mango is a cross of — **Dashehari × Neelam**

☆ Mallika variety of mango is a cross of — **Neelam × Dashehari**

- ☆ Ratna variety of mango is a cross of — **Neelam × Alphanso**
- ☆ Sindhu variety of mango is a cross of — **Ratna× Alphanso**
- ☆ Arka puneet variety of mango is a cross of — **Alphanso × Banganpalli**
- ☆ Arka aruna variety of mango is a cross of — **Banganpalli ×Alphanso**
- ☆ Pusa arunima variety of mango is a cross of — **Amarpali × Sensation**
- ☆ Mango malformation resistant varieties — **Illaichi, Bhaduran, Alib**
- ☆ Seedless variety of mango — **Sindhu**
- ☆ Most popular and exporter variety — **Alphanso**
- ☆ Suitable variety for processing — **Alphanso**
- ☆ Highly susceptible variety to spongy tissue — **Alphanso**
- ☆ Highly susceptible variety to malformation — **Bombay green**
- ☆ Dwarf variety of mango — **Amrapali**
- ☆ Row and planting spacing is generally kept for Amrapali is — **2.5 Meter**
- ☆ Sweetest variety of mango — **Chausa**
- ☆ Seli incompitable varieties — **Dashehri, Bombay green, Chausa**
- ☆ Polyembryonic Indian varieties — **Salem, Goa, Bellary, Olour**
- ☆ Suitable variety for canning — **Alphonso**
- ☆ Suitable variety for exporting — **Alphonso**
- ☆ Apple shaped mango variety — **Rumani**
- ☆ Off season mango variety — **Niranjan**
- ☆ Only mutant cultivar of mango — **Rosica**
- ☆ Regularity, malformation, spongy tissue, precocity and dwarfism are governed by — **Recessive gene**
- ☆ Major pest — **Mango hopper**
- ☆ Major disease — **Powdery mildew**
- ☆ Which variety of mango is most suitable for High Density Planting (HDP) — **Amarapali**
- ☆ In HDP 2.5×2.5 m² spacing for mango planting how many plant in planted in one ha. — **1600**
- ☆ Exported colored variety of mango — **Tommy atkins**
- ☆ Which variety of mango are free from spongy tissue — **Arka anmol, Arka puneet, Ratna**

☆ Regular bearing varieties — **Neelam, Ratna, Gulab khas, Tota pari, Banglora, Himsagar**

☆ Control of mango malformation can be done by — **Spray of NAA (200ppm) in Oct.-Nov. month**

Citrus

☆ Botanical name of mandarin — *Citrus reticulata*

☆ Botanical name of sweet orange — *Citrus sinensis*

☆ Botanical name of lemon — *Citrus lemon*

☆ Botanical name of acid lime/kagzi lime — *Citrus aurentifolia*

☆ Botanical name of sweet lime — *Citrus limetoides*

☆ Botanical name of grape fruit — *Citrus paradisi*

☆ Botanical name of Tahiti lime — *Citrus latifolia*

☆ Botanical name of rangpur lime — *Citrus limonica*

☆ Botanical name of pummelo — *Citrus grandis*

☆ Family — **Rutaceae**

☆ Fruit type — **Hesperidum**

☆ Propagation by — **Seed and Budding**

☆ Edible part — **Juicy placental hairs**

☆ Mandarin also called — **Fancy fruit**

☆ Grape fruit also called — **Break fast fruit**

☆ Normally planting space — **5×5m^2**

☆ Best irrigation method for citrus — **Ring method**

☆ Thornless species of citrus — **Tahiti lime**

☆ Most promising root stock of mandarin and sweet orange — **Rangpur lime**

☆ Only citrus fruit contain maleic acid — **Sweet orange**

☆ Seedless triploid citrus species — **Tahiti lime**

☆ Seedless variety of mandarin — **Satsuma**

☆ Seedless variety of lemon — **Lucknow**

☆ Mandrin is also known as — **Fancy fruit**

☆ Monoembryonic species of citrus — **Citron, Pummelo, Tahiti lime**

☆ Highly polyembryonic species of citrus — **Acid lime, grape fruit, Sweet orange, Mandrin**

☆ Sweet orange varieties — **Pineapple, Washington, Blood red malta**

☆ Grape fruit varieties — **Dunean, Foster, Saharanpur special**

☆ Kinnow variety of mandarin is a cross of — **King (*Citrus nobilis*) × Willow leaf (*Citrus deliciosa*)**

☆ Kinnow was developed by — **H.B. Frost at Citrus Experiment Station Riverside, California in 1935**

☆ First record of the name grape fruit in — **Jamaica (1814)**

☆ Kinnow mandarin was introduce in India — **1959**

☆ Citrus is a modified berry which is called — **Hesperidium**

Guava

☆ Botanical name — *Psidium guajava*

☆ Family — **Myrtaceae**

☆ Also known as — **Apple of Tropics**

☆ Propagation by — **Air layering/Stooling**

☆ Edible part — **Thalamus and Pericarp**

☆ Normal planting space — **10×10m²**

☆ Seedless ness in guava is due to — **Vegetative parthenocarpy**

☆ Flowering time of ambe bahar — **February – March**

☆ Fruiting time of ambey bahar — **July – September**

☆ Flowering time of mrig bahar — **June- July**

☆ Flowering time of hasth bahar — **October - November**

☆ Which time period is excellent for fruit quality — **Mrig bahar**

☆ Dual purpose variety — **Lalit**

☆ Parthenocarpy variety — **Allahabad round**

☆ Pusa srijan is the dwarfing root stock of — **Guava**

☆ Which guava variety is known as sardar — **Lucknow-49**

☆ Kohir safed variety of guava is a cross of — **Kohir × Allahabad safedah**

☆ Guava Hybrid-45 is the cross of — **Allahabadi safeda × L-45**

Papaya

☆ Botanical name of papaya — *Carica papaya*

☆ Virus resistant species of papaya is — *Carica cauliflora*

☆ Frost resistant species of papaya is — *Carica candamarcensis* and *Carica pentagona*

☆ Origin — Tropical America

☆ Commercially propagated by — Seed

☆ Seed rate — 500 g/ha for Dioecious and 250 g/ha for Gynodioecious

☆ Planting space — 2×2 m^2

☆ Yellow pigment in papaya is — Caricaxanthin

☆ Enzyme present in dried latex of papaya — Pepsin

☆ Carica papaya is — Polygamous

☆ Test weight — 50 g

☆ Highest papain yielding variety — Pusa majesty

☆ Dwarf mutant variety — Pusa Nanha

☆ Best variety for HDP (high density planting) 1.25×1.25 m^2 of papaya — Pusa Nanha

☆ Chemical used for better colour and keeping quality of papain — Potassium meta-bi-sulphite

☆ Gynodioecious varieties — Pusa majesty, Pusa delicious, Coorg honey dew, Co-3

☆ Dioecious varieties — Pusa Nanha, Pusa giant, Pusa dwarf, Co-6(selection from Pusa majesty)

☆ Major disease — Damping off

Banana

☆ Botanical name — *Musa paradisiaca*

☆ Family — Musaceae

☆ Also known as — Tree of paradise, Adam's fig

☆ Banana is a — Tropical herbaceous and monocotyledonous fruit

☆ Inflorescence of banana — Spadix

☆ Fruit type — Berries

☆ Propagation by — Sword suckers

☆ Normal planting space — 2×2m^2

☆ Degreening done by — Ethylene

☆ Lady finger is a variety of — **Banana**

☆ Removal of undesired suckers, done once in 45 days of planting is called — **Desuckering**

☆ Removal of male bud after completion of female phase is known as — **Denavelling**

☆ Tetrazolium test is used for detection of — **Bunchy top virus**

☆ Banana breeding was started in India (1949) at — **Central Banana Research Station, Aduthurai (T.N.)**

☆ Mutant varieties — **Robusta, Giant Cavendish, High gate**

☆ Inflorescence of banana consist of — **Female and hermaphrodite flowers**

☆ Banana initiate flowering — **9-12 months after planting**

☆ Seedless ness in banana is due to — **Vegetative parthenocarpy**

☆ Natural parthenocarpy is found in — **Banana**

☆ Group of carandis banana — **AAA**

Apple

☆ Botanical name of apple — *Malus domestica*

☆ Family — **Rosaceae**

☆ Propagation by — **Tongue grafting**

☆ Also called as — **King of temperate fruits**

☆ Edible part — **Fleshy Thalamus**

☆ Aroma in ripe fruit of apple is due to — **Iso-Pentanol**

☆ Acid found in apple is — **Malic acid**

☆ Redness in apple is due to — **Anthocyanin**

☆ Bitter pit disorder of apple is due to — **Calcium deficiency**

☆ Rosette disorder of apple is due to — **Zinc deficiency**

☆ Cider product prepare from — **Apple**

☆ M-27 variety of apple is a cross of — **M-13 × M-9**

☆ Ultra dwarf root stock is — **M-27**

☆ Triploid variety of apple is — **Baldwin**

☆ Diploid variety of apple is — **Self fertile**

☆ Wooly aphid resistant root stock — **Northern spy**

☆ Mother of all delicious group of variety is — **Red delicious**

☆ Appropriate leaf and fruit ratio found in Apple trees — 30:1

☆ Ratio of pollinizer and main crop in apple orchard should be — 9:1

☆ Sanjose scale is most serious pest of — **Apple**

☆ Which type of self incompatibility is found in apple — **Gemetophytic**

☆ Important varieties of apple — **Delicious, Sunheri, Chobatia princess, Red ambri**

☆ Regular bearing varieties of apple — **Rome beauty, Jonathan**

Important Points of other Fruit Crops

☆ Proper pit size for fruit crop — $1 \, m^3$

☆ Inflorescence of arecanut — **Spadix**

☆ Cashew nut is commonly known as — **Plough crop**

☆ name of pomegranate — *Punica granatum*

☆ Propagation of pomegranate by — **Air layering**

☆ Ganesh, Caveri and Dholka are varieties of — **Pomogranate**

☆ Juice of pomogranate is useful for patient suffering from — **Leprosy**

☆ Fruit cracking of pomegranate is most probable in — **Mrig bahar**

☆ Inflorescence of pear is — **Raceme**

☆ Propagation of pine apple by — **Suckers and slips**

☆ Giant knew, Mauritious, Singapore are variety of — **Pineapple**

☆ Which enzyme content present in pine apple — **Bromelin**

☆ Which chemical used for inducing flowering in pine apple — **NAA and Ethrel**

☆ Which chemical used for improving fruit quality of grape — **GA (20 ppm)**

☆ Arka hans is cross of — **Banglore blue × Anab-e-shahi**

☆ Fruit type of grape — **Berries**

☆ Propagation of grape by — **Hard wood cutting**

☆ Varieties of grape — **Pusa seedless, Pusa Navrang, Beauty seedless**

☆ Botanical name of ber — *Zyzyphus mauritiana*

☆ Commercially cultivated variety of ber — **Umran, Ganesh kirti**

☆ Early variety of ber — **Seb**

☆ Ber is commonly known as — **King of arid fruits and Poor man`s fruit**

☆ Best time for pruining of ber — **End of May to Mid June**

☆ Which crop is commonly known as single seeded nut — **Litchi**

☆ Fruit cracking in litchi is caused by — **Boron deficiency**

☆ Red pigment of lichi is due to — **Anthocyanin**

☆ Propagation of litchi by — **Air layering**

☆ Ne plus ultra and Sloh are variety of — **Almond**

☆ Amygdalin found in — **Almond**

☆ Propagation of jack fruit by — **Air layering, Seed**

☆ Marcottage is known as — **Air layering**

☆ Tea mosquito bug is a pest of — **Cashew**

☆ Botanical name of jamun — *Syzygium cumini*

☆ Family of jamun — **Myrteaceae**

☆ Which fruit recommended for sugar patient — **Jamun**

☆ Paras is a variety of — **Jamun**

☆ Narendra jamun -6 is a seed less variety of — **Jamun**

☆ Ampilography is cultivation of — **Vine crops**

☆ Colt is the root stock of — **Cherry**

☆ Commercially propagated of phalsa by — **Seed**

☆ Fruit type of phalsa — **Berries**

☆ Pusa early dwarf is varieties of — **strawberry**

☆ Harvesting time pf strawberry — **December - January**

☆ Botanical name of karonda — *Carica carandas*

☆ Family of karonda — **Apocynaceae**

☆ Propagation of aonla by — **Patch budding**

☆ Kanchan, Banarasi, Chakaiya, NA-7, Krishna varieties of — **Aonla**

☆ Most widely used training system of aonla — **Modified central leader system**

☆ Which fruit is best to control of pellagra disease — **Apricot**

☆ Plum blossoms aroma due to — **Benzaldehyde**

☆ Product of plum   – **Prunes**

☆ Myrobalan is the root stock of   – **Plum**

☆ Exanthema in citrus spp. occurs due to the lack of   – **Copper**

☆ Propagation of sapota by   – **Inarching**

☆ Cricket ball, Oval, Kali patti, Co-1 (Cricket ball × Oval) are variety of   – **Sapota**

☆ Manilkara root stock is used for which crop   – **Sapota**

☆ Propagation of date palm by   – **Off shoot**

☆ Botanical name of monkey jack   – ***Autocarpus heterophyllus***

☆ Champa, Rudrakshi, Singapore are variety of   – **Jack fruit**

☆ Marmelosin ingredient present in   – **Bael**

☆ Coffee is a   – **Short day plant**

☆ Family of coffee   – **Rubiaceae**

☆ Aroma in coffee is due to   – **Roastica**

☆ Kent is the mutant variety of   – **Coffee**

☆ Saffron is produced from   – **Styles and Stigma**

☆ Aroma in saffron due to   – **Safranal**

☆ Flavour in saffron is due to   – **Picrocrocin**

☆ Colour pigment in saffron is due to   – **Crocin**

☆ Term Bahar is related to   – **Guava and Pomegranate**

☆ Arka series of varieties are released from   – **IIHR, Bangalore**

☆ Which fruit is commonly known as Kalpvriksha   – **Coconut**

☆ Toddy is prepared by   – **Coconut**

☆ Origin of Rubber   – **Brazil**

☆ Tea is commercially propagated by   – **Soft wood cutting**

☆ Janam picking and Shear picking are done in   – **Tea**

☆ One bud and two leaf picking method is applied in   – **Tea**

☆ Food of god is known as   – **Cocoa**

☆ Most extensively fruit grown crop in the world   – **Mango**

☆ Cocoa fruit type   – **Drupe (Pod)**

☆ Cocoa bearing habit is   – **Cauliflorous**

☆ Rubber Research Institute situated in   – **Kottayam, Kerala**

- ☆ Custard apple and Avacado are — **Protogynous fruit**
- ☆ Richest source of Vit. A — **Mango after that papaya**
- ☆ Richest source of Vit. B1 — **Cashewnut after that walnut**
- ☆ Richest source of Vit. B2 — **Bael after that papaya**
- ☆ Richest source of Vit. C — **Barbados Cherry after that aonla**
- ☆ Richest source of Carbohydrate — **Apricot after that date**
- ☆ Richest source of Protein — **Cashewnut after that Almond**
- ☆ Richest source of Fat — **Walnut after that Almond**
- ☆ Richest source of fibre — **Guava after that Kaintha**
- ☆ Richest source of Calciun — **Litchi after that Karonda**
- ☆ Richest source of Iron — **Karonda after that Date**

II. Cultivation of Vegetable Crops

Potato

- ☆ Botanical name — *Solanum tuberosum*
- ☆ Family — **Solanaceae**
- ☆ Origin — **Peru (South America)**
- ☆ Potato contains large amount of the toxic alkaloid called — **Solanin**
- ☆ Solanin content present in potato is — **5 mg/100 g of potato**
- ☆ Protein content — **1.6 per cent**
- ☆ Most popular method for planting — **Ridge and furrow**
- ☆ Seed rate — **20-25 q/ha**
- ☆ Seed rate of True potato seed is — **100 g/ha**
- ☆ Seed plot technique was developed by — **Dr. Pushkernath**
- ☆ First Director of CPRI — **Dr. S. Ramanujan**
- ☆ True potato seed concept was first realize to raise commercial crop in India by — **Dr. S. Ramanujan**
- ☆ True potato seed (TPS) for planting one hectare — **40-45 g**
- ☆ True potato seed (TPS) concept was developed by — **Dr. S. Ramanujan**
- ☆ Earthing up done at — **30-45 DAS**
- ☆ Source of N_2 potato crop required fertilizer — **Potassium nitrate**

☆ Special size/superior grade tuber of potato
should have — **≥ 8 cm diameter**

☆ Research on potato was started at IARI, New Delhi in — **1935**

☆ ICAR-Central Potato Research Institute (CPRI) was
established at — **Shimla in 1949**

☆ Headquarter of CPRI was shifted from Patna to Shimla in — **1956**

☆ International Potato Centre (CIP) was established at — **Lima (Peru) in 1971**

☆ In India potato was introduce by Portugese in — **1615**

☆ Most serious disease of potato — **Late blight**

☆ Most critical stage for irrigation — **25 per cent tuber
formation stage**

☆ Desuckering of potato is done by — **Malaic Hydrazide
(2 per cent)**

☆ Storage in ventilated closed room with — **4-5^{0}C temperature and
90-95 per cent RH**

☆ varieties of True potato seed of potato — **C3, HPS 1/13, HPS 2/67**

☆ Suitable variety for processing and producing
light colour chips and finger fries — **Kufri chipsona-1**

☆ Dehulming of potato is used to — **Obtain quality seed tuber
by using the chemical CuSO$_4$**

☆ Short duration varieties — **Kufri chandramukhi, Kufri alankar,
Kufri jyoti, Kufri bahar**

☆ Suitable for late planting varieties — **Kufri jeevan, Kufri sinduri,
Kufri dewa**

☆ Selected Tuber for sowing should have — **at least 3 buds, diameter 3cm
with 30 g weight**

☆ Seed plot technique (SPT) in potato used for — **Producing virus free
seed tubers**

☆ Potato tubers are modified form of — **Stem**

☆ Potatoes are born on — **Stolons**

Chilli

☆ Botanical name — ***Capsicum annum***

☆ Fruit type — **Berries**

☆ Red colour in chilli due to — **Capsenthin**

☆ Pungency of chilli due to — **Capsicin**

☆ Chemical use for fruit setting — Triacontanol

☆ Suitable variety for HDP — Jwalamukhi

☆ Leaf curl resistant varieties — **Pusa jwala, Pusa sadabahar**

☆ Green : Dry ratio in chilli — 10:1

☆ Pungency of chilli measured by — **Scoville scale**

☆ Oleoresin is an important product of — **Chilli**

Tomato

☆ Botanical name — *Solanum lycopersicum* **(Old - *Lycopersicon esculentium*)**

☆ Family — **Solanaceae**

☆ Origin — **Peru**

☆ Also known as — Wolf apple

☆ Fruit type — **Berry**

☆ Pigment responsible for red colour — **Lycopene**

☆ Lycopene is highest at — $21^0C–25^0C$

☆ Acidity of tomato is due to — **Citric acid and Malic acid**

☆ No.1 processing vegetable is — Tomato

☆ Best method of extraction of tomato seed — **Alkali method**

☆ Tomato is Susceptible for — Frost

☆ Blossom end rot (BER) disease is due to — **Ca deficiency**

☆ Fruit cracking of tomato is due to — **B deficiency**

☆ Most important nutrient required for tomato cultivation — **B and Zn**

☆ Seed rate — **300-350 g**

☆ Pusa ruby is a cross of — **Sioux × Improved maruti**

☆ Best variety for drought condition — **Arka vikas**

☆ Suitable variety for low temperature — **Pusa Sheetal**

☆ Suitable variety for high temperature — **Pusa Hybrid-1**

☆ Suitable variety for low as well as high temperature — **Pusa Sadabahar**

☆ First root knot nematode resistant variety — **Sel-120**

☆ Nematode and Bacterial wilt resistant variety — **Arka vardan**

☆ Chemical used for tomato sauce preservation — **Sodium benzoate**

☆ Suitable varieties for processing — **Arka Saurabh, Pusa Gaurav, Pusa Uphar, Roma**

☆ Varieties — **Pant bahar, Pusa ruby, Arka sourabh, Ashish, Arka vikas, Arka vardan**

☆ Major pest — **Fruit borer and root knot nematode**

Cabbage

☆ Botanical name — ***Brassica oleraceae* var. *capitata***

☆ Seed rate — **350 – 500g/ha**

☆ Edible part — **Head**

☆ Black rot resistant variety — **Pusa Mukta**

☆ Anti cancer property of cabbage is due to — **Indole – 3- Cardinal**

☆ Varieties — **September, Pride of India**

Cauliflower

☆ Botanical name — ***Brassica oleraceae* var. *botrytis***

☆ Edible part — **Curd**

☆ Blanching is Important practices of — **Cauliflower**

☆ Seed rate — **500-600 g/ha**

☆ Varieties — **Pusa snowball, Early Kunwari, Pusa Deepali, Pusa Shubhra**

☆ Whip tail in cauliflower is due to — **Mo deficiency**

☆ Browning in cauliflower is due to — **B**

☆ Blanching is an important process of — **Cauliflower**

Knol Khol

☆ Botanical name — ***Brassica oleraceae* var. *gongylodes***

☆ Also known — **Swollen stem (Knol-rabi)**

☆ Seed rate — **1-1.5 kg/ha**

☆ Economic part — **Extended stem**

Brinjal

☆ Botanical name — ***Solanum melongena***

☆ Also known as — **Egg plant, Aubergine**

☆ Seed rate — **200 – 250 g/ha**

☆ Fruit type — **Berries**

☆ Good soure of — **Vitamin B**

☆ Brinjal flowers are 4 type — **Long Styled, Medium Styled, Short Styled, Pseudo Short Styled**

☆ White brinjal is preferred by — **Diabetic patient**

☆ Extra early maturing variety — **Pusa purple long**

☆ Bacterial blight and Phomosis blight resistant variety — **Pant samrat**

Okra

☆ Fruit type — **Capsule**

☆ Seed rate — **8 – 10 kg/ha**

☆ Varieties — **Parbhani kranti, Pusa sawani**

☆ Most serious disease — **Yellow vein mosaic**

☆ YMV resistant variety — **Pusa sawani**

Onion

☆ Pungency due to — **Allyl propile disulphide**

☆ Temperature required for bolting in onion is — **<15⁰C**

☆ For seed production purpose, onion is considered as — **Biennial plant**

☆ Irritation of eye due to cutting onion is due to presence of — **Pyruvic acid or Sulphur compound**

☆ Chemical used to prevent sprouting during storage of onion is — **Maleic hydrazide (MH)**

Water Melon

☆ Seed rate — **3.5 – 5 kg/ha**

☆ How much water contain by water melon — **95 per cent**

☆ Sugar baby is the popular variety of — **watermelon**

☆ Seed less F1 hybrid — **Pusa Bedana**

☆ Pusa bedana is a cross of — **Tetra-2 × Pusa Rasal**

☆ Pigment present in water melon for different colour — **Anthocynin and Lycopene**

Pumpkin

☆ Botanical name — ***Cucurbita moschata***

☆ Seed rate — **1 – 1.5 kg/ha**

☆ Chief Pollinator — **Honey bees**

Bottle gourd

- ☆ Botanical name — *Lagenaria siceraria*
- ☆ Seed rate — 3 – 4 kg/ha
- ☆ Flower colour — **White**
- ☆ Varieties — **Prolific long, Pusa manjari, Pusa summer, Pusa meghdoot, Pusa naveen**

Cucumber

- ☆ Gynomonoecious flower found in — **Cucumber**
- ☆ Fruit type — **Pepo**
- ☆ Bitterness in cucumber due to — **Metaxenia**
- ☆ First F_1 hybrid variety of Cucumber is — **Pusa sanyog**

Pointed Gourd (Parwal)

- ☆ Botanical name — *Trichosanthes dioica*
- ☆ Propagation by — **Vine/Stem cutting**

Important Points of other Vegetable Crops

- ☆ Which is vary good source of Iron, Calcium, Vit-C, Vit-K, Carotenoid is — **Kale**
- ☆ Pusa chikni, Harita, Pusa supriya is variety of — **Sponge gourd**
- ☆ Most suitable harvesting stage of muskmelon is — **Full slip stage**
- ☆ Satputia, Pusa Nasdar is variety of — **Ridge gourd**
- ☆ Arka Suryamukhi is variety of — **Summer squash**
- ☆ Toxic substance of cucurbits — **Cucurbitacins**
- ☆ Vegetative parthenocarpy is found in — **Kundru**
- ☆ Which vegetable fruit is used for making Petha — **Wax gourd/Ash gourd**
- ☆ Ivy gourd also known as — **Little gourd, Coccinia**
- ☆ Which growth regulator is isolated from Yam — **Batasin**
- ☆ Only tuber crop, which is rich in protein — **Colocasia**
- ☆ Toxic substance present in colocasia — **Calcium Oxalate**
- ☆ Product of cassava is — **Sago**
- ☆ Chow-Chow is propagated by — **Sprouted fruits**
- ☆ Economic part of sweet potato — **Adventious root**

☆ Tomato, Brinjal, Cowpea, Beans are — **Day Neutral Vegetable plant**

☆ First hybrid variety in vegetable developed in — **Bottle gourd**

☆ Richest source of protein — **Beans**

☆ Term Bareja system is associated with — **Betelvine**

☆ Which harmones are used in cucurbits to modify sex and to induce femaleness — **NAA and GA**

☆ Round gourd (Tinda) also known as — **Indian squash, Squash melon**

☆ Cabbage, Cauliflower, Brussels sprout, Kale, Sprouting broccoli vegetable group is — **Cole crop**

☆ Alkaloid VINBLASTIN is extract from — **Periwinkle**

☆ Which spice crops is largely exported from India — **Black pepper**

☆ Pungency in garlic due to — **Allinase**

☆ Pusa meghali is variety of — **Carrot**

☆ Richest source of Vit. A — **Carrot**

☆ For seed purpose carrot is grown as — **Biennial**

☆ Seed rate of Bitter bourd — **4.5 – 5 kg/ha**

☆ Botanical name of Brussels sprouts is — *Brassica oleraceae* **var.** *gemmifera*

☆ Botanical name of Sprouting broccoli — *Brassica oleraceae* **var.** *italica*

☆ Botanical name kale is — *Brassica oleraceae* **var.** *acephala*

☆ Japanese white is variety of — **Raddish**

☆ Diara cultivation method is followed in — **Cucurbits**

☆ Sulphur is not applied in which crop — **Cucurbits**

III. Cultivation of Flowers

Rose

☆ Botanical name — *Rosa indica*

☆ Family — **Rosaceae**

☆ Rose name come from — **Latin word rosa**

☆ Also called — **King of flowers**

☆ Propagation by — **T budding**

☆ Rose are ancient symbol of — **Love and Beauty**

☆ White rose can be used to convey — **Sympathy or Humanity**

☆ Yellow rose can be used to convey — **Friendship and caring**

☆ Red rose can be used to convey — **Love, respect, and devotion**

☆ Deep red rose can be used to convey — **Heartfelt regret and Sorrow**

☆ Pink rose can be used to convey — **Joy and gratitude**

☆ Orange rose can be used to convey — **Passion and energy**

☆ Lavender rose can be used to convey — **Love at first sight**

☆ Black rose can be used to convey — **Death of feelings or Idea**

☆ Fruit of rose is called — **Hip**

☆ Rose hip have very high content — **Vitamin C**

☆ Propagation by — **T-budding or Shield budding**

☆ Most popular root stock for North India — ***Rosa indica, Rosa barboniana***

☆ Most suitable root stock for South India — ***Rosa multiflora***

☆ First hybrid Tea rose is — **La France**

☆ Dainty, Ice berg, Tuscan are varieties of — **Floribunda rose**

☆ Red fairy, Pink fairy are varieties of — **Polyantha rose**

☆ Blaze is variety of — **Climber rose**

☆ Thornless variety of rose — **Chitra**

☆ Queen Elizabeth is a variety of — **Floribundas rose**

☆ Queen Elizabeth and Montezuma are variety of — **Grandiflora rose**

☆ Gulkand is prepared by mixing rose petals and sugar in ratio of — **1:1**

☆ From two thousand flowes produce how much oil — **1g**

☆ Famous rose breeder in India is — **Dr. B.P. Pal**

☆ Father of rose breeding — **Dr. Bhatta chaterji**

☆ Leading exporter of rose in worlds — **Netherland**

☆ $C_{10}H_{18}O$ is chemical formula of — **Geraniol**

☆ Mohini cultivar of rose is famous for its — **Chocolate flower colour**

☆ Floribunda rose is a cross of — **Hybrid tea × Dwarf polyantha**

☆ Main constituent of attar of rose are — **Geraniol, I-citronelllol and Fragrant alcohols**

Carnation

- ☆ Commercially propagated by — **Terminal cutting**
- ☆ Carnation is — **Herbaceous perennial plant**
- ☆ Dark red carnation can be used to convey — **Deep love and affection**

Chrysanthemum

- ☆ Botanical name — *Chrysanthemum indicum*
- ☆ Family — **Asteraceae**
- ☆ Chrysanthemum word come from — **Greek word chrysous means Golden**
- ☆ Also known as — **Glory of East**
- ☆ Growth habit — **Perennial**
- ☆ Flower colour — **White and Yellow**
- ☆ Crown bud in chrysanthemum appers in month of — **May**
- ☆ Propagation by — **Root suckers and terminal cutting**
- ☆ Off season varieties — **Jaya, Jwala, Haldi ghati**

Gerbera

- ☆ Botanical name — *Gerbera jamesoni*
- ☆ Origin — **Mediterranean region**
- ☆ Gerbera contains — **Coumarin derivatives**
- ☆ 5th most used cut flower in the World after Rose, Carnation, Chrysanthemum, Tulip is — **Gerbera**

Jasmine

- ☆ Botanical name — *Jasminum grandiflorum*
- ☆ Family — **Oleaceae**
- ☆ *Jasminum sambae* is called as — **Mogra**
- ☆ *Jasminum arborescens* is called as — **Naba mallika**
- ☆ *Jasminum humile* is called as — **Swarn chameli/Italian jasmine**
- ☆ Jasminum auricultum is called as — **Joohee**
- ☆ Jasmine flower called in Hindi word — **Chameli**

Gladioulus

- ☆ Botanical name   –
- ☆ Family   – **Iridaceae**
- ☆ Origin   – **South Africa**
- ☆ Also known as   – **Sword lily**
- ☆ Gladiolus is a symbol of   – **Power and Strength**
- ☆ Propagation by   – **Corm**
- ☆ How much corms required for one hectare   – **2.0 to 2.5 lakh corms**
- ☆ The word gladiolus derived from   – **Latin word gladius mean sword**
- ☆ Important variety of gladiolus are   – **Palampur Queen, Tushar, Grace, Nanus, Oscar, Mayur, Pusa Suhagin, Melody, White Prosperity, Red Beauty, Jester Gold, Nova Lux, Suchitra**

Rajani Gandha

- ☆ Scientific name   – *Poloanthes tuberosa*
- ☆ Family   – **Amaryllidaceae**
- ☆ Origin   – **Maxico**
- ☆ Also known as   – **Tuberose**
- ☆ Propagation by   – **Tuber**
- ☆ Double type of tuberose is known as   – **Pearl**
- ☆ Tuberose is known in South India   – **Sugandhraja (King of the Fragrance)**
- ☆ Tuberose is known in Tamil Nadu   – **Nilasambangi or Sambangi**
- ☆ Tuberose is known in Bengal   – **Rojoni Gondha (Scent of the night)**
- ☆ Tuberose is known in Iran   – **Mary flower**
- ☆ Rajni Gandha means   – **Night fragrant (Rajni means Night and Gandha means Fragrance)**
- ☆ Polianthes means   – **Grey flower**
- ☆ Important variety of Rajni Gandha   – **Rajat Rekha, Pearl Double, Maxican single**

Orchid

- ☆ Botanical name — *Vanilla planifolia*
- ☆ Flower is — **Cross pollination type**
- ☆ Vanillin is extracted from — **Orchid**
- ☆ Orchids grown in two pattern like — **Monopodial and Sympodial**

Bougainvillea

- ☆ Family — **Nyctaginaceae**
- ☆ Origin — **South America**
- ☆ Bougainvillea glabra is some times called as — **Paper flower**
- ☆ R.R. Pal is root stock of — **Bougainvillea**

Merigold

- ☆ Botanical name of African Marigold — *Tagetes erecta*
- ☆ Botanical name of French Merigold — *Tagetes petula*
- ☆ Patels of Tagetes are rich in orange yellow carotenoid called — **Lutein**
- ☆ Seed rate — **1-1.5kg.ha**

Calendula

- ☆ Botanical name — *Calendula officinalis*
- ☆ Calendula is also known as — **Pot Marigold/True merigold**

Important Points

- ☆ National flower of India — **Lotus**
- ☆ National flower of Netherlands — **Tulip**
- ☆ National flower of USA — **Rose**
- ☆ National flower of Bulgaria — **Rose**
- ☆ National flower of Switzerland — **Edelweiss**
- ☆ National flower of Norway — *Silver saxifrages*
- ☆ National flower of China — **Narcisuss**
- ☆ National flower of Japan — **Chrysanthemum**
- ☆ National flower of England — **Rose**
- ☆ Bonsai culture originated from — **Japan**
- ☆ Growing of tree in very dwarf form (5-20 cm) is known as — **Mame Bonsai**

☆ Concept of lawn was developed in — **England**

☆ Indian botanical garden is a example of — **English style garden**

☆ Largest producer of cut flower in the world is — **Netherland**

☆ Which country produced maximum rose oil — **Bulgaria**

☆ Baradari is an important feature of — **Mughal garden**

☆ Mughal garden "Nishat" was constructed by — **Asaf Khan (Brother of Noorjahan w/o Jahangir)**

☆ Style of pinjor garden — **Formal or Mughal**

☆ Lal bagh garden is situated at — **Bangalore**

☆ Biggest rock garden in India situated at — **Chandigarh**

☆ Floral clock is an important feature of — **Lal Bagh garden**

☆ Ram bagh garden is situated in Agra was construced by — **Babar**

☆ Taj garden situated in — **Agra**

☆ Municipal park situated in — **Massorrie**

☆ Pijor garden situated in — **Pinjor**

☆ Motibagh garden situated in — **Patiala**

☆ Nasimbagh and Shalimar garden situated in — **Kashmir**

☆ Talkatora garden situated in — **New Delhi**

☆ Arrangement of rocks is an important features of — **Japanese garden**

☆ Term Heaven of Man is used for the — **Italian garden**

☆ Kamla Nehru park in Mumbai is famous for — **Topiary**

☆ Cottage green concept was given by — **G- Jakyell**

☆ Largest flower is — **Rafflesia**

☆ Smallest flower is — **Wolffia**

☆ Smallest seed is — **Orchid**

☆ Largest grass is — **Bamboosa**

☆ Largest consumptive loose flower is — **Marigold**

☆ Crown of Gold and Golden Age are variety of — **African Marigold**

☆ Who first started tissue culture in orchids — **Morel**

☆ World's No.1 foliage plant is — **Dieffenbachia**

- ☆ Herkogamy favour to — **Cross pollination**
- ☆ Oldest banyan tree is located at — **Indian botanical garden, Kolkata**
- ☆ Avenue tree related to Lord Krishna is — *Anthocephalus Cadamba*
- ☆ Shrubs or tree planted at regular intervals on boundary for fencing is called — **Hedges**
- ☆ Fastest method of lawn making — **Turfing**
- ☆ Most ancient type of garden is — **Kitchen garden**
- ☆ Garden city of India is also known as — **Bengaluru**
- ☆ Xeriscaping is a — **Arid land scaping**
- ☆ Second shilajeet is called — **Safed musli**
- ☆ Soil erosion in orchard can be checked by — **Sod culture**
- ☆ Best time for orchard soil culture is — **November - January**
- ☆ Planting of low growing plants along with paths, roads, flowers beds, lawn, etc. for demarcation and beautification is called — **Edge**
- ☆ The art of developing the plant or training the plant into different forms or shapes like animals, birds, etc. is called — **Topiary**
- ☆ Family of antirrhinum — **Plantaginaceae**
- ☆ Antirrhinum is also known as — **Snap dragon**
- ☆ Family of lily — **Liliaceae**
- ☆ Propagation of lily by — **Bulb**
- ☆ Botanical name of china aster — *Callistephus chinensis*
- ☆ Cultivation of grapes is called — **Viticulture**
- ☆ Making wine from grapes is called — **Viriculture**

IV. Post-Harvest Technology

- ☆ CSIR-Central Food Technological Research Institute (CFTRI) is situated at — **Mysore**
- ☆ Father of food preservation — **Nicolas Apart**
- ☆ TSS of cooking jam is measured by — **Hand Refractometer**
- ☆ Pectin content is measured by — **Jellimeter**
- ☆ Solution made by dissolving salt (NaOH) in water is called as — **Brine solution**
- ☆ Which fruit beverage is diluted before serving — **Syrup**

☆ Mango pulp is preserved by — **Sugar**

☆ TSS of jam should not be — **<70 per cent**

☆ Concentration of salt sufficient to preserve most of the food products is — **15-25 per cent**

☆ Concentration of sugar required for preservation of fruits and jam is — **66 per cent**

☆ Salt concentration in pickle is maintained at — **8-10 per cent**

☆ Semi solid transparent product prepared from pectin containing fruit is — **Jelly**

☆ pH of final jelly should be — **3.2**

☆ TSS of Tomato sauce is — **30**

☆ Permissible limit of SO_2 in jam is — **40 ppm**

☆ Juices are mostly preserved by — **Freezing**

☆ Wine is preserved at — **14 per cent alcohol**

☆ Ideal storage temperature for papaya — **9-11°C**

☆ Acetic acid present in — **Vinegar**

☆ Which is a class one preservative — **Sugar**

☆ Original colour of beverages for longer period are retained by — **Benzoic Acid**

☆ A heat treatment food material at 72°C for 15 minutes or 63°C for 30 minutes called — **Pasteurization**

10
Plant Physiology

☆ Father of plant physiology   **– Stephen Hales**

☆ Root pressure word given by   **– Stephen Hales**

☆ Osmosis concept given by   **– Abbe Nollet (1748)**

☆ Imbibition term was coined by   **– Sachs**

☆ Mass flow or pressure flow theory was given by   **– Godlewski (1884)**

☆ Cohesion and Adhesion theory or Transpiration pull theory was given by   **– Dixon and Joly (1894)**

☆ Auxin term was first used by   **– Frits Warmolt Went (1905)**

☆ Photoperiodism term was coined by   **– Garner and Allard (1920)**

☆ Vernalisation term was coined by   **– Lysenko (1920)**

☆ Pulsation theory was given by   **– J. C. Bose (1923)**

☆ Gibberellin was discovered by   **– Kurosawa (1926)**

☆ Diffusion pressure deficit (DPD) term was introduced by   **– Meyer (1938)**

☆ Phytohormone for hormone of plant term is suggested by   **– Thimann (1948)**

☆ Who traced the path of carbon in photosynthesis and gave C_3 cycle   **– Calvin (1954)**

☆ Who establish that ethylene is the only gaseous growth regulator — **Burg (1962)**

☆ Abscisic acid was first identified by — **Addicott (1963)**

☆ Cytokinin term proposed by — **Letham (1963)**

☆ Abscisic acid first identified by — **Wareing (1965)**

☆ Zeatin was first identified by — **D. S. Letham (1974)**

☆ Photorespiration term was first used by — **Deeker (in Tobacco)**

☆ Law of limiting factor explained by — **Blackman (1905)**

☆ Leaf area indeax by — **Watson (1947)**

☆ Kinetin was identified by — **Miller and Skoog**

☆ Lock and key model was proposed by — **Fisher**

☆ Hormone related to drought tolerance is — **Abscisic acid**

☆ Irreversible change in any plant part with respect to size, form, weight, volume is called — **Growth**

☆ Phasic change of individual cells into tissues, organs, and organisms is called — **Development**

☆ Particle moves from higher to lower concentration region is called — **Diffusion**

☆ Movement of water takes place from lower to higher concentration of solution is called — **Osmosis**

☆ Which sugar is known as fruit sugar — **Fructose**

☆ Which sugar is known as cane sugar — **Sucrose**

☆ Which sugar is known as universal sugar — **Glucose**

☆ Which sugar is known as milk sugar — **Lactose**

☆ Lactose is a combination of — **Glucose + Galactose**

☆ Most important sugar acid is — **Ascorbic acid**

☆ Plasma membrane is type of — **Semi permeable membrane**

☆ Diffusion of solvent particles into a living cell is called — **Endosmosis**

☆ Strong solution having higher concentration — **Hypertonic solution**

☆ Shrinkage of protoplasm due to outward flow of water (exosmosis) in concentrate solution — **Plasmolysis**

☆ Casparian strip is present in — **Endosmosis**

☆ Hydrostatic pressure generated within the cell against cell wall as a result of entry of water into it, due to osmosis — **Turgor pressure**

☆ Water is absorption by plants mainly through — **Root hairs**

☆ Loss of water from the injured parts of the plant is called — **Bleeding**

☆ Hydrostatic pressure developed due to the accumulation of water absorb by the root is called — **Root pressure**

☆ Most accepted theory of water absorption — **Transpiration Pull theory**

☆ Imbibition pressure is also known as — **Matric potential**

☆ Tricontanol is a growth stimulant obtain from — **Leaves of Lucerne**

☆ First step in absorption of water is — **Imbibition**

☆ Absorbed of water against a concentration gradient by using energy released from respiration is called as — **Active absorption**

☆ Transpiration associated ion uptake is — **Passive absorption**

☆ Nutrients absorbed by plants from soil solution are carried upward through — **Xylem**

☆ Downward movement of food synthesized in leaves takes place through — **Phloem**

☆ Movement of nutrient ions and salts along with moving water — **Mass flow**

☆ Upward translocation of fluid in xylem takes place due to — **Pull of transpiration stream**

☆ Plant cell wall are — **Permeable in nature**

☆ Minerals are translocated in plants as — **Organic and inorganic compounds**

☆ Plant meet their carbon requirement by absorbing — **CO_2 from atmosphere**

☆ Process by which plants convert light energy of photon (capture from sunrays) into chemical energy — **Photosynthesis**

☆ Oxidation reduction process is — **Photosynthesis**

☆ Carotene and Xanthophyll, fat sluble yellow pigment are called — **Carotenoids**

☆ Phycobilins pigment as found in — **Blue and Red algae**

☆ Absorption spectrum of photosynthesis are — **Blue and Red**

☆ Action spectrum of photosynthesis are — **Red and Blue**

☆ Material with surface that bind water is called — **Matrix**

☆ Photosynthesis active radiation (PAR) occurs at — **400-700 nm**

☆ Ca is essential for — **Cellwall formation**

☆ Chlorophylls occur mostly in the grana and associated with — **Thylakoids membrane**

☆ Plant component responsible for Photosynthesis is a pigment called — **Chlorophyll**

☆ Chlorophyll contains — **Mg**

☆ Chemical formula of chlorophyll a — $C_{55}H_{72}O_5N_4Mg$

☆ Chemical formula of chlorophyll b — $C_{55}H_{70}O_6N_4Mg$

☆ First biological process that begins in a seed soon after in imbibes water — **Respiration**

☆ One $NADH_2$ produces how many ATP molecules — **3**

☆ Red hemoglobin in root nodule is responsible for — **Transport of oxygen**

Photosynthesis

☆ Photosynthesis is an — **Oxidation-Reduction process**

☆ Who reported C_3 pathway for CO_2 fixation in certain tropical grasses — **Hatch and Slake (1965)**

☆ Reduction of CO_2 to carbohydrate level needs assimilatory product such as — **ATP and $NADH_{+2}$**

☆ One $NADH_2$ will produce — **3 ATP**

☆ One $FADH_2$ will produce — **2 ATP**

☆ Reduction of CO_2 occurs in — **Dark**

☆ Grana and Stroma are found in — **Chloroplast**

☆ In photosynthesis light energy converted into — **Chemical energy**

☆ Important accessory pigments in plant are — **Carotenoids and Xanthophylls**

☆ Pigment which are responsible for Photosynthesis in higher plants — **Chlorophyll a and b**

☆ Colour of chlorophyll a — **Blue green**

☆ Colour of chlorophyll b — **Yellow green**

☆ Oxygen required for photosynthesis comes from — **Water**

☆ Product of photosynthesis which is used for growth and development of plants — **Glucose**

☆ Photosynthesis can be measured by measuring — **O_2 given off and CO_2 uptake**

☆ Chemical which retard transpiration rate called — **Anti-transparent**

☆ Which organism do not have photosynthesizing capability — **Fungi**

☆ Photorespiration is also called as — **C_2 cycle**

☆ Photorespiration is common in — **C_3 plants**

☆ Dark reaction was discovered by — **Black man**

☆ C_3 cycle also called — **Dark reaction or Calvin cycle or Black man reaction**

☆ Dark reaction take place in — **Stroma of chlorophyll**

☆ CO_2 acceptor in C_3 cycle is — **Ribulose-1,5-diphosphate (RUBP)**

☆ First stable product C_3 cycle is — **Phosphoglyceric acid (3 PGA)**

☆ In C_3 cycle for synthesis of one glucose molecule ATP required are — **18**

☆ Rubisco or RUBP carboxylase enzyme found in — **Stroma**

☆ Most abundant protein on earth — **Rubisco or RUBP carboxylase**

☆ C_3 plant are — **Paddy, Wheat, Barley, Pea, Soybean, Pulse, Cotton, Soybean**

☆ Most important enzyme involved in photosynthetic CO_2 fixation in C_3 plants — **Rubisco**

☆ Water use efficiency of C_4 is high than — **C_3 plants**

☆ Photosynthetic rate of C_4 is low than — **C_3 plants**

☆ Light reaction was discovered by — **Hatch and Slake (1970)**

☆ C_4 cycle also called — **Light reaction or Hill reaction or Hatch and Slake pathway**

☆ Light reaction take place in — **Grana of chlorophyll**

☆ CO_2 acceptor in C_4 cycle is — **Phosphoenol pyruvic Acid (PEP)**

☆ First product of C_4 pathway is — **Oxalo Acetic Acid (OAA)**

☆ In C_4 cycle for synthesis of one glucose molecule ATP required are — **30**

☆ Photosynthetic rate is highest in — **C_4 plants**

☆ PEPCO enzyme present in — **C_4 plants**

☆ C_4 plants are — **Maize, Sugarcane, Sorghum, Millets**

☆ Most important enzyme involved in photosynthetic CO_2 fixation in C_4 plants — **PEP carboxylase**

☆ C_4 plants have — **Kranz type leaf**

☆ C_4 plants normally give more biological yield than C_3 plants because of — **Less respiration**

☆ Product of light reaction — **ATP and NADPH$_2$**

☆ CAM denotes — **Crassulacean Acid Metabolism**

☆ CAM cycle occurs in — **Mesophylls cells**

☆ CAM plants are — **Pineapple, Opuntia, Bryophyllum, Agave**

☆ Water use efficiency are — **CAM > C_4 > C_3**

☆ CO_2 concentration in atmosphere is — **382 ppm**

☆ Photosynthesis reaction — **$6CO_2 + 12H_2O + Sun\ light \rightarrow C_6H_{12}O_6 + 6H_2O + 6O_2$**

Respiration

☆ Oxygen is required by the plants for — **Respiration**

☆ Respiration in plant consists of — **Glycolysis, Krebs cycle, Electron Transport Chain (ETC)**

☆ Glycolysis also called — **Embden Meyer of Paranas Pathway (EMP Pathway)**

☆ Glycolysis occurs in — **Cytoplasm**

☆ Glycolysis occurs in — **Anaerobic condition**

☆ First product of glycolysis — **Glucose-6-phosphate**

☆ Glycolysis involve degradation of glucose to — **Pyruvate**

☆ Final product of glycolysis — **Pyruvate**

☆ Net gain ATP synthesis from one molecule of glucose in glycolysis — **2**

☆ Gross production ATP synthesis from one molecule of glucose in glycolysis — **4**

☆ Anaerobic respiration pathway product are — **Ethanol and lactic acid**

☆ Krebs cycle, Electron Transport Chain occurs in — **Mitochondria**

☆ Krebs cycle occurs in — **Matrix of Mitochondria in aerobic condition**

☆ Krebs cycle also called — **Citric Acid Cycle (CAC), Tricarboxylic Acid Cycle (TAC)**

☆ Krebs cycle starts with — **Acetyl Co-A and Oxaloacetate**

☆ Krebs cycle produces — **30 ATP**

☆ ETC also called — **Respiratory chain or oxidative phosphorylation**

☆ ETC occurs in — **Inner membrane of cristae in mitochondria**

☆ ETC is present in — **Cristae of mitochondria (where ATP is synthesized in respiration)**

☆ Electron carriers involved in the respiratory ETC — **Cytochromes**

☆ Net gain ATP synthesis from one molecule of glucose in respiration — **36 ATP**

☆ Gross production ATP synthesis from one molecule of glucose in respiration — **38 ATP**

☆ No of CO_2 molecule released between anaerobic and aerobic respiration is — **Zero**

☆ Ratio of energy released between anaerobic and aerobic respiration is — **1:18**

☆ Ratio of photosynthesis to respiration during day time is — **10:1**

☆ Ratio of CO_2 reduced and oxygen released during photosynthesis is — **1:1**

☆ One molecules of ATP is equal to — **7.6 K cal energy**

☆ One molecule of $NADH_2$ is equal to — **52 K cal energy**

☆ One molecule of glucose is produce energy — **686 K cal**

☆ Energy currency in the cell is — **ATP**

☆ Ratio of CO_2 evolved to ratio of O_2 evolved in plant is called — **Respiratory Quotient**

☆ Respiratory Quotient (RQ) of carbohydrate — **1**

☆ Respiratory Quotient (RQ) of Protein — **0.8**

☆ Respiratory Quotient (RQ) of fat — **0.7**

☆ An energy spending process — **Photorespiration**

☆ Photorespiration occurs in — **Night**

☆ Photorespiration is highest in — **Paddy**

☆ In photorespiration NAD is reduced to — **$NADH_2$**

☆ Photorespiration occurs only in — **Chlorophyllous cells**

☆ Loss of water in the form of vapour from the living aerial parts of the plant is known as — **Transpiration**

☆ Principle organ of transpiration is — **Stomata of leaf (90 per cent transpiration)**

☆ Water is lost during transpiration in the form of — **Vapour**

☆ Lose of water through hydathodes in liquid from during night and regulated by root pressure — **Guttation**

☆ Transpiration occurs in — **Day**

☆ Guttation occurs in — **Night**

☆ Stomata is found mainly on — **Lower surface of leaves**

☆ Opening and closing of stomata are due to its — **Turgidity and Flaccidity**

☆ Element which takes part in the growth and development of plants — **Plant nutrients**

☆ Essential element for opening and closing of stomata — **K^+**

☆ Type of stomata mostly present on lower surface of leaves — **Potato type**

☆ Stomata that is present on only under surface of leaf — **Apple and Mulberry type**

☆ Oat type stomata found in — **Equally distributed on both surface**

Plant Growth Hormone (PGR's)

☆ Organic substance which are manually produced in plant are — **Phytohormone**

☆ Hormone also known as — **Growth hormone, Growth substance, Phytohormone**

☆ Organic compounds which inhibit or modify any physiological process — **Plant growth regulator (PGR)**

☆ Growth promoters are — **Auxin, Gibberellins and Cytokinin**

☆ Growth inhibitors are — **Abscisic acid, Ethylene**

☆ Growth retardant is — **Cycocel (CCC)**

☆ Auxin word derived from — **Greek word**

☆ Auxin hormones are — **IAA, IBA, 2,4-D, MCPA, 2,4,5-T, TIBA, IPA etc.**

☆ Widely used auxin herbicide are — **2,4-D, MCPA, 2,4,5-T**

☆ Main auxin is also known as — **IAA**

☆ Natural auxin is — **IAA**

☆ Synthetic produced auxin are — **NAA, IBA, 2,4-D, MCPA**

☆ Root formation — **IAA, IBA**

☆ Enhance fruit set and number of fruits — **NAA**

☆ Anti auxin are — **Naphthythalamic acid (NTA), Ethylene chlorohydrins**

☆ Site of auxin transport is located on — **Plasmalemma**

☆ Test that are generally used for bioassays of auxin — **Avena curvature test and Split pea stem curvature**

☆ Apical bud dominance is caused by — **Auxin**

☆ Active sites of auxin — **Shoot tip region, Coleoptile and Developing embryos**

☆ Auxin synthesis occurs rapidly in — **Green leaves in presence of light than in the dark**

☆ Growth hormone which is weakly acidic in nature — **Auxin**

☆ Went's auxin is now known as — **Indole-3-acetic acid (IAA)**

☆ Which herbicide used as a harmone — **2,4-D (< 20 ppm)**

☆ 2,4-d stand for — **Dichlorophenoxy acetic acid**

☆ Movement of gibberellin take place in — **Xylem and Phloem**

☆ Chemically gibberellin are related to — **Terpenoids**

☆ Precursor of gibberellin — **N-Kaurene**

☆ Gibberellin first isolated from — *Gibberella fujikuroi*

☆ Cycocel, Maleic hydrazide, Paclobutrazol, Phosphon-D are — **Anti Gibberellin**

☆ Gibberellin were first invent in — **Japan**

☆ Site of cytokinin synthesis — **Root tip**

☆ Cytokinin is mainly synthesized in — **Root tips**

☆ Movement of cytokinin take place through — **Xylem**

☆ First naturally occurring hormone identified — **Zeatin**

☆ Senescence is delayed by — **Cytokinin**

☆ Seed dormancy breaking and stimulate cell division hormone is — **Cytokinin**

☆ Precursor of cytokinin — **Adenine**

☆ Mobility of cytokinin — **Polar and Basipetal**

☆ Naturally occurring growth regulator is — **Abscisic acid**

☆ PGR acting as stress hormone — **Abscisic acid**

☆ Seed dormancy is induced by — **Abscisic acid (ABA)**

☆ Abscisic acid is synthesized from — **Actively growing points**

☆ PGR related to drought tolerance and stress hardness in plants — **Abscisic acid (ABA)**

☆ ABA is called — **Anti-gibberellins**

☆ Shedding of plant parts is due to — **ABA**

☆ Precursor for biosynthesis of Abcisic acid — **Violoxanthin**

☆ Biosynthesis of Abcisic acid also takes place through — **Mevalonic acid**

☆ Chemically ABA are related to — **Terpenoids**

☆ Bioassays for ABA are — **Paddy seedling growth inhibition test and inhibition of amylase in barley endosperm**

☆ Destruction product of violoxanthin and forms ABA — **Xanthoxin**

☆ Precursor of IAA is — **Tryptophane**

☆ Synthesis of IAA in plants requires — **Zinc**

☆ Ethylene is a volatile gas which is included under hormone in — **1971**

☆ Ethylene is synthesized in plant from — **Methionine amino acid**

☆ Ethylene level in plant increase by — **Auxin**

☆ Merely gaseous hormone is — **Ethylene**

☆ Which PGR used for fruit ripening — **Ethylene**

☆ Naturally occurring volatile hormone — **Ethylene**

☆ Maximum ethylene is formed in — **Ripening fruits and senescing tissue**

☆ Biosynthesis of ethylene occurs from — **Methionine**

☆ Inhibitor of ethylene synthesis — **Amino-ethoxyvinylglycine**

☆ Bioassays for ethylene test — **Triple pea test and pea stem swelling test**

☆ Root promoting hormone is — **IBA**

☆ Formation of male flowers is induced by — **GA$_3$**

☆ Physiological response of plants in relation to length of light — **Photoperiodism**

☆ Short day plants required day length — **<10 hrs**

☆ Paddy, Potato, Tobacco, Sugarcane, Soybean, Onion, Strawberry, Chrysanthemum and generally kharif crops are — **Short day plants**

☆ Kharif crops required — **Shorter day length**

☆ Generally rabi crops are — **Long day plants**

☆ Long day plants required day length — **>14 hrs**

☆ Wheat, Barley, Castor, Oat, Alfalfa, Spinach, Radish, Sugar beet, Opium poppy and generally rabi crops are — **Long day plants**

☆ Cotton, Pea, Maize, Sunflower, Tomato, Cucumber are — **Day neutral plant**

☆ Method on inducing early flowering in plants by pre treatment of their seeds at very low temperature is called — **Vernalisation**

☆ Which hormone responsible for Vernalisation — **Vernalin**

☆ Sites of vernalisation — **Apical buds/ Growing points**

☆ One atm is equal to — **1.01 bars**

☆ Soil less cultivation of plant is known as — **Hydroponics**

☆ Root pressure is measured by — **Manometer**

☆ Growth rate of plant is measured by — **Auxanometer and Crescograph**

☆ Respiratory Quotient (RQ) of a plant material measured by — **Ganong`s Respirometer**

☆ Rate of transpiration is measured by — **Potometer**

Important Points Related to PGRs

☆ Foolish seedling of paddy caused by — *Gibberella fujikuroi*

☆ Foolish seedling of paddy is commonly called — **Bakanae disease of paddy**

☆ Hormone used for sugarcane ripening — **Glysophosine**

☆ Hormone used for seed less grape — **GA$_3$**

☆ Hormone used for flower initiator and fruit setting — **NAA**

☆ Hormone used for sucker control in tobacco — **MH**

☆ Hormone used for ripening of fruit — **Ethylene**

☆ Wound hormone found in injured portion of plant — **Traumatic acid**

☆ Aflatoxin is produced by — *Aspergillus flavus*

☆ Plant age determined by — **Annual rings**

☆ Element that play an important role in the protoplasm constituent is — **Calcium**

☆ Highest affinity of haemoglobin with which gas — **CO**

☆ Steroid isolated from pollen grains of brassica is — **Brassins**

☆ Methyl ester in jasmine, inhibits growth and promote senescence — **Jasmonic acid**

☆ Lichens are symbiotic relation between — **Algae and Fungus**

☆ Non cyclic photo phosphorylation consist of — **PS$_{700}$**

☆ Phenyl mercuric acetate (PMA) chemical used in agricultural crop for — **Reduce tranpiration**

Role of PGRs

Sl.No.	Plant Growth Hormone	Role
1.	Auxin	Promotes apical dominance
		☆ Promotes elongation of cells
		☆ Increase cell division in cambium
		☆ Induce uniform flowering in pineapple
		☆ Increase in shoot and decrease in root formation
		☆ IBA promotes rooting of cutting
2.	Gibberellin	Breaking the dormancy
		☆ Promote male flower production
		☆ Enhance seed germination
		☆ Induction of flowering in long day plants
		☆ Stem elongation
3.	Cytokinin	Initiation of cell division
		☆ Delay of senescence
		☆ Induce flowering in short day plant
		☆ Promote stomata opening
		☆ Promote femaleness in male flower
4.	Ethylene	Fruit ripening with increase in respiration
		☆ Induce uniform flowering and ripening in pineapple
		☆ Inhibit stem elongation and cause abscission of leaves
		☆ Induce fruiting in ornamental plants
		☆ Latex flow increase in rubber
5.	Abscissic acid	Induce bud dormancy and enhance process of abscission
		☆ Stimulate release of ethylene
		☆ Senescence of leaf is promoted
		☆ Bring closure of stomata during water stress

11
Genetics

☆ Simple microscope was discovered by — **Galileo (1610)**

☆ First compound microscope was made by — **Robert Hooke (1665)**

☆ Cell was discovered by — **Robert Hooke (1665)**

☆ Nucleus was discovered by — **Robert Brown (1983)**

☆ Cell theory was given by — **M. J. Schleiden and T. Schwann (1939)**

☆ Nucleolus was discovered by — **Fontona (1781) and it was described by Wagner (1840)**

☆ Theory of evolution through natural selection was given by — **Darwin and Wallace (1858)**

☆ Concept of pangenesis was developed by — **Darwin**

☆ Theory of acquired character was developed by — **Lamarck**

☆ Genomice term was coined by — **Thomas Roderick (1986)**

☆ Ergastoplasm term was given by — **Garnier (1900)**

☆ Lysosome term was first used by — **Duve (1955)**

☆ Mitochondria term was given by — **Benda (1897)**

☆ Mitochondria was discovered by — **Hollicker**

☆ Endoplasmic reticulum term was given by — **Porter (1948)**

☆ Plastids was introduced by — **Lederberg**

☆ Germ plasm term was first used by — **Weismann**

☆ Genetic resource term was coined by — **Frankel**

☆ Genic balance system of sex determination
was proposed by — **Bridge (1922)**

☆ Parthenogenesis term was coined by — **Owen**

☆ Chromosomal theory of heredity was proposed by — **Sutton (1902)**

☆ Chromosomal theory of inheritance was proposed by — **Sutton and Boveri (1903)**

☆ Gene, Genotype, Phenotype term used by — **Johnson (1903)**

☆ Chromosome term coined by — **Waldeyer (1888)**

☆ Chromosome was firstly discovered by — **Strasburger (1875)**

☆ Gemete term was first used by — **Strasburger (1877)**

☆ Genetics term coined by — **W. Betson (1905)**

☆ Father of modern genetics — **G. J. Mendel**

☆ Law of heredity was first discovered by — **Mendel**

☆ One gene one enzyme hypothesis was given by — **Beadle and Tatum (1941)**

☆ Chiasma type theory of crossing over was proposed by — **Janssen**

☆ Gene for gene hypothesis is known as — **Flor hypothesis**

☆ Centre of origin was first given by — **Vavilov**

☆ How many centre of origin recognized by Vavilov — **8**

☆ How many kind of cells are found in living world — **2 (Prokaryote and Eukaryote)**

☆ Plant cell is a type of cell — **Eukaryote**

☆ Cell wall of plant cell are composed of carbohydrate
known as — **Cellulose**

☆ Largest organelle in Eukaryotic cell — **Nucleus**

☆ Functional unit of life in known as — **Cell**

☆ Which organelle of cell is non living — **Cell wall**

☆ Material contained in vacuoles — **Cell sap**

☆ Physical basis of life is known as — **Protoplasm**

☆ Cells without cell wall is called — **Protoplasts**

☆ Controlling centre of cell is — **Nucleus**

☆ Nucleus contain — **Genetic material**

☆	Mitochondria is also called	**– Power house of cell**
☆	Name the prokaryotic organism which does not contain mitochondria	**– Bacteria**
☆	Which cell organelles is found in both prokaryote and eukaryote cells	**– Ribosome**
☆	70s type ribosome is found in	**– Mitochondria**
☆	Ribosome is also called	**– Engine of cell**
☆	Main site of Protein synthesis is	**– Ribosome**
☆	Ribosome of 80s size found in	**– Plant and Animal**
☆	Ribosome of 70s size found in	**– Prokaryote and Eukaryote**
☆	Main function of golgi body is	**– Packaging of food materials**
☆	Cytoplasm and Nucleus combinedly called	**– Protoplasm**
☆	Endoplasmic reticulum with attached ribosome is called	**– Rough Endoplasmic reticulum (RER)**
☆	Endoplasmic reticulum for	**– Lipid synthesis**
☆	Lysosome is also known as	**– Suicidal bag of cell**
☆	Damage cell is digest by	**– Lysosome**
☆	Main function of golgi body	**– Packing and transportation of food material**
☆	Colourless plastid is called	**– Leucoplast**
☆	Vacuoles is also known as	**– Dustbin of cell**
☆	Self replicating, extra chromosomal genetic material found in plant cell	**– Plastids**
☆	Which organelles of cell are found only in plants	**– Plastid and Spherosome**
☆	Which plastid of cell is responsible for photosynthesis in plants	**– Chloroplast**
☆	Chlorophyll synthesis occur in	**– Chloroplast**
☆	Which plastid of cell is responsible for colour in plants	**– Chromoplast**
☆	Which plastid of cell is responsible for storage of starch and fat in plants	**– Leucoplast**
☆	Name of leucoplast which functions as the storage of oil	**– Lipoplast**
☆	Thread like bodies that carry gene	**– Chromosome**
☆	Fundamental unit of chromosome	**– Chromatin**

- ☆ Which part of chromosome is known as primary constriction — **Centromere**
- ☆ Major genetic constituent of chromosome — **DNA**
- ☆ DNA is a polymer of — **Nucleotides**
- ☆ DNA core of nucleosome is called — **Octasome**
- ☆ Double helix model of DNA proposed by — **Watson and Crick (1953)**
- ☆ New cells which are formed by the process of cell division is called — **Daughter cell**
- ☆ Division of chromosomes and cytoplasm of a cell into two daughter cells is known as — **Cell division**
- ☆ Stage in spindle using cell division during which DNA synthesis takes place is called — **Interphase**
- ☆ Interphase is also known as — **Resting phase**
- ☆ Substage of Interphase — **G_1, S, G_2 phase**
- ☆ Interphase term first used by — **Lundergardh (1912)**
- ☆ Interphase is — **Pre DNA replication phase**
- ☆ Post DNA replication phase during which protein and RNA synthesis takes place is — **G_2 phase**
- ☆ Prophase, Metaphase and Anaphase terms was coined by — **Strasburger (1884)**
- ☆ Telophase term was coined by — **Heidenhain (1894)**
- ☆ Karyokinesis term was first used by — **Schleicher (1878)**
- ☆ Equational division term was first used by — **Weismann (1887)**
- ☆ Cytokinesis tern was coined by — **Weismann (1887)**
- ☆ Process of division of cytoplasm is called — **Cytokinesis**
- ☆ Process of division of nucleus is called — **Karyokinesis**
- ☆ Stage of DNA synthesis in mitosis — **Interphase**
- ☆ Mitosis term was coined by — **W. Flemming (1882)**
- ☆ Mitosis occurs at — **Somatic cells**
- ☆ Mitosis division also called — **Equational division**
- ☆ Main function of mitosis — **Growth of organism and regeneration of damaged tissue**
- ☆ Sequence of mitosis phase — **Prophase → Metaphase → Anaphase → Telophase**

☆ Longest phase of mitosis is — **Prophase**

☆ Longest phase of cell cycle is — **Interphase**

☆ Shorted phase of mitosis is — **Anaphase**

☆ Middle stage in which chromosome are arrange in equatorial plate — **Metaphase**

☆ How many daughter cells are formed in one cycle of mitosis — **2**

☆ Nucleolus disappears at — **End of Prophase**

☆ Nucleolus reappears at — **End of Telophase**

☆ Meiosis term was first given by — **Farmer and Moore (1905)**

☆ Meiosis cell division was first described by — **Beneden (1883)**

☆ Sequence of Meiosis stage — **Leptotene→ Zygotene → Pachytene → Diplotene→ Dikinesis**

☆ Meiosis division also known as — **Reduction division**

☆ Meiosis occur at — **Reproductive cells**

☆ In plants, meiosis take place in — **Anther and Ovary**

☆ Segregation occurs during — **Meiosis**

☆ Bivalent and Crossing over take place in which stage of meiosis — **Pachytene stage**

☆ Chiasmata occur at — **Diplotene**

☆ Spindle formation take place during — **Metaphase-I**

☆ Process of separation of chromatids called — **Disjunction**

☆ How many daughter cells are formed in one cycle of meiosis — **4**

☆ Meiosis having — **Two cytokinesis**

☆ Mitosis having — **One cytokinesis**

☆ Lampbrush chromosomes is visible during — **Diplotene**

☆ Tendency of two of more genes to remain together in the same chromosome during inheritance is referred to as — **Linkage**

☆ Physical location of a particular gene on a chromosome is called — **Locus**

☆ Mendel was born on — **22 July 1822**

☆ Mendel died in — **1884**

☆ Mendal paper entitled "Experiments in plant hybridization" was presented in — **German language**

☆ Mendal work on the 7 contrasting character of which crop — **Pea**

☆ Rediscovery of Mendel work was done by — **Hugo de Vries, Erich Correns, Erich Tschermak (1900)**

☆ Accepted theory of Mendel was — **Law of Segregation**

☆ Mendelian population is also known as — **Random mating population**

☆ Many alleles of a single gene are called — **Multiple alleles**

☆ Alternative form of a gene is known as — **Allele**

☆ Tall pea plant (TT) and tall pea plant (Tt) have what in common — **Phenotype**

☆ Trait which appeared in F1 generation was called — **Dominant**

☆ Exons and introns term are related to — **Split genes**

☆ Expression of one gene depends on the presence or absence of another gene in an individual — **Gene interaction**

☆ Interaction between alleles of two or more different loci is known as — **Epistasis**

☆ Phenotype ratio of mono cross hybrid — **3:1**

☆ Phenotype ration of di cross hybrid — **9:3:3:1**

☆ Phenotypic F_2 ratio in Complementary gene action is — **9:7**

☆ Phenotypic F_2 ratio in Duplicate gene action is — **15:1**

☆ Phenotypic F_2 ratio in Supplementary gene action is — **9:3:4**

☆ Phenotypic F_2 ratio in Inhibitory gene action is — **13:3**

☆ 5 inbred lines will lead How many number of single crosses — **10**

☆ Process of protein synthesis — **DNA —transcription→ m-RNA —translation→ Protein**

☆ Initiation of protein synthesis requires — **mRNA**

☆ Triplet sequence found in mRNA which codes for single amino acid — **Codon**

☆ Primary codon are — **AUG**

☆ Termination codon are — **UAA, UAG, UGA**

☆ Triplet sequence in t-RNA — **Anticodon**

☆ Sex chromosome also known as — **Allosomes**

☆ Genes for sex linked traits are located on — **Y- chromosome**

☆ How many pairs of homologous chromosome do humans have — **23**

☆ Ploidy level in embryo — **2n**

☆ Ploidy level in endosperm — **3n**

☆ Ploidy level in testa — **2n**

☆ Ploidy level in aleuren — **2n**

☆ Base + Sugar is — **Nucleoside**

☆ Base + Sugar + PO_4 is — **Nucleotide**

☆ Progeny test was developed by — **Louis de vilmorin**

☆ Okazaki fragments are produced during — **DNA replication**

☆ Replication of DNA is — **Bidirectional**

☆ One helicle turn of DNA helix measures — **3.4 mm**

☆ Most useful insect well studied in genetics is — ***Drosophila melanogaster***

☆ ELISA plate is made up of — **Polyvinyl chloride or Polystyrene**

☆ ELISA test is used to detect the presence of — **HIV**

☆ External protein coat of virus particle is called — **Capsid**

12
Plant Breeding

- ☆ Primary centres of origin was proposed by — **Vavilov**
- ☆ Heterosis term was given by — **Shull**
- ☆ Crossing over term was first time used by — **Morgan and Cattell (1912)**
- ☆ Father of hybrid cotton — **C. T. Patel**
- ☆ First hybrid of paddy was developed by — **Yuan Longping**
- ☆ First transgenic plant was developed by — **Fraley in tobacco (1983)**
- ☆ Operon model of gene regulation was discovered by — **Jacob and Monod (1961)**
- ☆ Pass pedigree method was proposed by — **Harrington (1937)**
- ☆ Progeny test was developed by — **Louis de Vilmorin**
- ☆ Single seed descent method applied for the first time in oat by — **Grafius (1965)**
- ☆ Recurrent selection term was coined by — **Hull (1945)**
- ☆ Use of synthetic varieties for commercial cultivation was first suggested in maize by — **Hayes and Garber (1919)**
- ☆ Vertical and horizontal resistance concept developed by — **Van der Plank (1963)**
- ☆ Dominance theory was proposed by — **Davenport (1908)**
- ☆ Theory of central dogma was proposed by — **Crick**

☆ NBPGR (National Bureau of Plant Genetic Resource) situated at — **New Delhi**

☆ NBPGR (National Bureau of Plant Genetic Resource) established in — **1976**

☆ Vegetative embryos develops without fertilization is called — **Apomixis**

☆ Development of fruit without fertilization is called — **Parthenocarpy**

☆ Embryo originates from unfertilized egg is called — **Parthenogenisis**

☆ Progeny of a single cross fertilized heterozygous individual — **Inbred**

☆ Progeny of a single plant obtained by asexual reproduction — **Clone**

☆ Single gene affecting more than on character is called — **Pleiotropy**

☆ Mechanism of self pollination in which flowers open but only after pollination has take place — **Chasmogamy**

☆ Chasmogamy was found in — **Paddy, Oat, Moong**

☆ Pollination and fertilization occurs before opening of flower is termed as — **Cleistogamy**

☆ Cleistogamy was found in — **Wheat, Barley**

☆ When male and female flowers of a hermaphrodite flower matures at different time — **Dichogamy**

☆ When female matures before mane — **Protogyny**

☆ Pollen from a flower of one plant falls on the stigma of other flowers of the same plant — **Geitonogamy**

☆ When male and female flowers occur on the same plant — **Monoecious**

☆ When male and female flower occur on different plant — **Dioecious**

☆ Example of monoecius plant is — **Maize**

☆ Example of dioecoius plant is — **Papaya**

☆ Synthetic varieties are maintained by — **Open pollination**

☆ Synthetic variety can be developed from — **Inbreds, clones and open pollination varieties**

☆ Basic concept in the development of synthetic varieties is — **Exploitation of hybrid vigour**

☆ When an F_1 hybrid is cross with one of its parent — **Back cross**

☆ Cross of F_1 parent with homozygous recessive parent — **Test cross**

☆ Intra specific hybridization is a crosses between — **Two plant of different varieties**

- ☆ When pollen grains from an another falls on receptive stigma of the same flower — **Self pollination**

- ☆ Self pollination species are also known as — **Autogamous species**

- ☆ When pollen grains from flower of one plant transferred to receptive stigas of flowers of another plant — **Cross pollination**

- ☆ Continuous inbreeding (selfing) leads — **Homozygosity**

- ☆ Single seed descent method is a method of — **Self pollination**

- ☆ Mass selection is always based on — **Phenotype**

- ☆ Oldest selection method of crop improvement — **Mass selection**

- ☆ Maximum variability — **Mass selected variety**

- ☆ Concept of pure line theory was given by — **Johnson**

- ☆ In pureline theory, Johnson was working on — **Princess variety of Rajma (*Phaseolus vulgaris*)**

- ☆ Lines are homozygous and homogenous in nature is called — **Pure line**

- ☆ Homozygous Progeny of a self pollinated homozygous plants is called — **Pure line**

- ☆ Method of breeding is not appropriate for cross pollinated crops — **Pure line selection**

- ☆ Method of breeding for wheat — **Pure line selection**

- ☆ Bulk method was first used by — **Nilsson Ehle (1908)**

- ☆ Breeding refers to selection procedure in which the segregation population of self pollinated species is grown without selections — **Bulk**

- ☆ Method which is not for handling segregation populations — **Bulk method**

- ☆ Method for improving specific traits *i.e.* plant height, disease — **Pedigree method**

- ☆ Most commonly used method for selection from segregation generations of crosses in self pollinated crops — **Pedigree method**

- ☆ Parent which is used only once in back cross breeding method — **Donor**

- ☆ Progeny selection is also known as — **Ear to row method**

- ☆ Methods used for handling the segregation generation — **Pedigree, Bulk and Single seed descent method**

☆ Modified of bulk method is — **Single seed descent method**

☆ Multi line breeding is exploited widely in the crop — **Wheat**

☆ Method does not provide opportunity to practice selection for superior plant till F_5 generation — **Single seed descent method**

☆ Commonly used method for transfer of disease resistancy from one variety to another variety — **Back cross method**

☆ Method of breeding is appropriate for improvement of good variety — **Back cross method**

☆ Method in which desirable scattered favourable genes are selected in different plants in each generation — **Recurrent selection**

☆ Cross between two genetically different homozygote plants is — **Hybrid or F_1**

☆ Single cross hybrid — **$A \times B = F_1$**

☆ Double cross hybrid — **$(A \times B) \times (C \times D) = F_1$**

☆ Selected variety/line/clone × open pollinated variety — **Top cross**

☆ Single cross (A×B) × open pollinated variety — **Double top cross**

☆ Variety produced by crossing in all combinations a number of lines that combine well each other — **Synthetic variety**

☆ Synthetic variety is maintained by — **Self pollination**

☆ Variety produced by mixing the seeds of several phenotypically outstanding lines/varieties — **Composite variety**

☆ Composite variety is developed by — **Cross pollination**

☆ Hybrid variety was first exploited in — **Maize**

☆ First intergeneric hybrid was — ***Raphinobrassica* (Radish × Cabbage)**

☆ Hybrid variety of paddy is developed by using — **GMS and CGMS line**

☆ Double cross hybrids of maize are developed by — **CGMS line**

☆ Exploitation of hybrids in tobacco was carried out by — **Koelreuter**

☆ Superiority of F_1 hybrids over both of its parents is termed as — **Heterosis**

☆ Heterosis is also known as — **Hybrid vigour**

☆ Average value for a character of two parents of the concerned hybrid — **Average heterosis**

☆ Superiority of F_1 over the standard commercial check variety is called — **Useful heterosis**

☆ When heterosis estimated over the superior or better
parent is called — **Heterobeltiosis**

☆ When superiority of hybrid to the standard commercial
check variety — **Economic heterosis**

☆ Exchange of chromatin between non-sister chromatids
of homologous chromosome is known as — **Crossing over**

☆ Two lines different for a single locus called — **Isogenic line**

☆ Hybridization between two such individuals of the
pedigree of a parent is called — **Inbreeding**

☆ Loss or decrease in vigour and fitness as a result of
inbreeding — **Inbreeding Depression**

☆ Removing immature anthers from any bisexual flower
is called — **Emasculation**

☆ Highly ID is found in — **Alfalfa and Carrot**

☆ Sudden heritable change in any characterstics of
an organism — **Mutation**

☆ X – ray as mutagen was first used by — **Muller**

☆ Unit in which mutation occurs — **Muton**

☆ Chemical or physical agent which greatly enhances the
frequency of mutation — **Mutagen**

☆ Man made cereal — **Triticale**

☆ Progeny test is also known as — **Vilmorin isolation principle**

☆ Removal of the entire tassel (male inflorescence of maize)
from the plant before pollen to initiate cross hybridization — **Detasseling**

☆ Condition in which either pollen is absent or non
functional in flowering plants — **Male sterility**

☆ When pollen sterility is caused by cytoplasmic genes — **Cytoplasmic male
sterility (CMS)**

☆ Effect of pollen on endosperm expression is called — **Xenia**

☆ Which crop is called drosophila of crop plants — **Maize**

☆ Ethrel is used as gametocide for — **Wheat, Paddy, Sugarbeet**

☆ Allohexaploidy found in — **Common bread wheat
(*Triticum aestivum*)**

☆ Autopoly ploidy found in — **Sugarcane, Cotton, Brassica,
Sweet potato, Oat, Alfalfa**

- ☆ Autotetraploidy found in — **Potato, Groundnut, Coffee, Alfalfa**
- ☆ Allopolyploids fonds in — **Wheat, cotton, tobacco and oat**
- ☆ Allotetraploids found in — **Cotton, Tobacco**
- ☆ Autotriploidy found in — **Banana**
- ☆ Triploid found in — **Apple, Watermelon, Sugarbeet**
- ☆ *Brassica nigra* was evolved from — ***Brassica compestris* × *Brassica oleracia***
- ☆ *Brassica juncia* was evolved from — ***Brassica compestris* × *Brassica nigra***
- ☆ Characteristic which are governed by several genes each having small individual effect — **Polygenic traits**
- ☆ Natural genetic engineer — ***Agrobacterium tumefaciens***
- ☆ Which soil born bacterium is used for developed of transgenic plants — ***Agrobacterium tumefaciens***
- ☆ Clonal selection mostly used in the crop — **Ginger**
- ☆ Male sterile line — **A- line**
- ☆ Basic chromosome number is — **X**
- ☆ Genetic chromosome number is — **n**
- ☆ Haploid no. of wheat — **n=21**
- ☆ Change in genome with reference to individual chromosomes called — **Aneuploidy**
- ☆ Heteroploid in which one or few chromosomes or missing from 2n — **Aneuploidy**
- ☆ Monosomic hypoploid — **2n-1**
- ☆ Double monosomic — **2n-1-1**
- ☆ Nullisomic — **2n-2**
- ☆ Hyperploid have one extra chromosome — **Trisomic (2n+1)**
- ☆ Double trisomic — **2n+1+1**
- ☆ Trtrasomic — **2n+2**
- ☆ Double tetrasomic — **2n+2**
- ☆ Monotrisomy — **2n-1+1**
- ☆ Polyploidy level in endosperm of wheat — **63**
- ☆ Triploids are useful for — **Seed less fruit**

- ☆ Paddy, Wheat, Oat are — **Self pollinated crops**
- ☆ Maize pearl millet, Mustard, Sunflower are — **Cross pollination crops**
- ☆ Safflower, Arhar, Cotton, Sorghum, Pea are — **Often cross pollination crops**
- ☆ Development of seed by self pollination is called — **Autogamy**
- ☆ Jagannath is a mutant variety of — **Paddy**
- ☆ Resistance of a host to the particular race of a pathogen is known as — **Vertical resistance**
- ☆ Vertical resistance is governed by — **One or few gene**
- ☆ Vertical resistance is also called as — **Oligogenic resistance**
- ☆ Resistance of a host to all the races of a pathogen is called — **Horizontal resistance**
- ☆ Disease resistance is governed by — **Several genes**
- ☆ Gene for gene hypothesis is also called — **Flor hypothesis**
- ☆ When resistance is governed by single major gene is known as — **Monogenic resistance**
- ☆ Dwarfing gene in paddy — **Dee-Gee-Woo-Gene**
- ☆ Dwarfing gene in wheat — **Norin-10**
- ☆ Exotic variety of paddy — **TN-1**
- ☆ Exotic variety of wheat — **Sonara-64 and Lerma rojo**
- ☆ Gregg 399 is an important source of GMS in — **Cotton**
- ☆ Kabir 60 is source of CMS is — **Sorghum**
- ☆ Tift 23A is source of CMS is — **Pearl millet**
- ☆ Non traditional area of paddy cultivation — **Punjab**
- ☆ Non traditional area of wheat cultivation — **West Bengal**
- ☆ Cereals are deficient in — **Lysine**
- ☆ Pulses are deficient in — **Tryptophan, Methionine**
- ☆ First hybrid of Paddy — **APHR-1**
- ☆ First interspecific hybrid of Cotton — **H4**
- ☆ First interspecific hybrid of Barley — **RBD-1**
- ☆ First interspecific hybrid of Bajra — **HB**
- ☆ Single gene dwarf varieties of wheat — **Sonalika, UP-262, WL-711, Girija**
- ☆ Double gene dwarf varieties of wheat — **Kalyansona, Arjun**

☆ Triple gene dwarf varieties of wheat – **Hira moti, Jyoti, Sangam, Jawahar**

☆ Karnal bunt resistant variety – **HD-2643**

☆ All rust resistant variety – **NP-804**

☆ Protein rich maize varieties – **Shakti, Rattan, Protina**

☆ Highest area of transgenic crop – **Soybean**

☆ First ideotype plant was developed in – **Wheat**

☆ First improved wheat variety in India – **HD-2285**

☆ World's first high yielding aromatic paddy – **Pusa Basmati**

☆ Pusa Basmati-1 is cross between – **Pusa-167 × Karnal**

☆ First hybrid of Pigeon pea – **ICPH-8**

☆ First hybrid paddy in India – **KRH-2**

☆ Allogamous crop is protogynous in nature – **Pearl millet**

☆ Zero and double zero are the varieties in – **Mustard**

☆ Most method of producing deletion mutant is – **X-ray irradiation**

☆ Plant phenomena responsible for checking the natural out crossing is – **Karyogamy**

☆ Karyogamy takes place in – **Aceiospore**

☆ First transgenic maize was developed in – **USA**

13
Seed Technology

☆ Fertilized ovule consisting of intact embryo, stored food and seed coat which is viable and has got the capacity to germinate — **Seed**

☆ National Seed Corporation (NSC) was established on — **March 1963**

☆ National Seed Corporation (NSC) was started functioning in — **July 1963**

☆ Seed Act was passed on — **1966**

☆ Seed Rule was passed on — **1968**

☆ National Seed Project (NSP) was started on — **1988**

☆ Headquarter of National Seed Corporation (NSC) — **New Delhi**

☆ Central Seed Testing Laboratory is situated — **NSRTC, Varanasi**

☆ National Seed Research and Training Centre (NSRTC) — **2005**

☆ State Seed Corporation in India — **13**

☆ State Seed Certification Agencies — **19**

☆ Source of breeder seed — **Nucleus seed**

☆ Progeny of breeder seed — **Foundation seed**

☆ Progeny of foundation seed — **Certified seed**

☆ Mother seed is also known as — **Breeder seed**

☆ Tag colour of Nucleus seed — **No tag**

☆ Tag colour of Breeder seed — **Golden yellow colour**

☆ Tag colour of Foundation seed — **White colour**

☆ Tag colour of Certified seed — **Blue colour**

☆ Tag colour of Registered seed — **Purple colour**

☆ Seed produced by NSC — **Foundation seed**

☆ Certified seed is generally produced by — **SSC**

☆ Which type of germination is found in Cereals, Gram, Arhar, Lentil — **Hypogeal**

☆ Which type of germination is found in Mustard, Sunflower, Castor, Onion — **Epigeal**

☆ Varietal purity is checked by — **Grow out test (GOT)**

☆ Impurity percentage of seed is also referred to as — **Dockage**

☆ Weight of 1000 seeds is called — **Test weight**

☆ Weight of 100 seed is called — **Seed index**

☆ Seed processing is termed as — **Grading**

☆ Standard method of seed moisture estimation — **Oven dry method**

☆ Dormancy due to hard seed coat or impermeable seed coats — **Scarification**

☆ Dormancy due to low temperature and moisture condition — **Stratification**

☆ PGR used to initiate seed germination — **Gibberellic acid**

☆ Seed dormancy of potato tubers is broken by treating tuber with — **Thiourea 1 per cent**

☆ Paddy variety which has no seed dormancy — **IR-50**

☆ Cold test is a type of vigour test used for — **Maize**

☆ Photolytic seed is found in — **Tobacco**

☆ Recurrent selection was coined by — **Hull (1945)**

☆ Bulk method was first used by — **Nilsson Ehle (1908)**

☆ Widely used fungal bio-control agent for seed treatment is — ***Trichoderma viride***

☆ Paddy and Wheat is — **Self pollinated crop**

☆ Most common method for improvement of self pollination crops is — **Pure line selection**

☆ Which class of seed is not generally used in India — **Registered seed**

☆ Formula of Real value of seed — =**Purity per cent × Germination per cent/100**

☆ Formula of Pure live seed — =**Purity per cent × Viability per cent/100**

☆ Capacity of the seed to germinate — **Seed vigour or Viability**

☆ Seed viability/Vigor is mostly tested by — **2,3,5- triphenyl tetrazolium chloride**

☆ Seed testing refers to — **Testing of purity, moisture, germination of seeds**

☆ Sequence of seed — **Nucleus seed→ Breeder seed→ Foundation seed→ Certified seed**

☆ Seed of a crop variety produced by the breeder which is small in quantity is called — **Nucleus seed**

☆ Main objective of field inspection is to examine — **Disease incidence, Isolation distance, Off type Plant**

☆ Seed fail to germinate under unfavourable environment is termed as — **Quiescent seed**

☆ Apomixis is related to — **Onion and Citrus**

☆ Development of embryo without fertilization is called — **Apomixis**

☆ Development of fruits without fertilization is called — **Parthenocarpy**

☆ Heterosis is a — **Hybrid vigour**

☆ Male sterility is used for the production of — **Hybrid varieties**

☆ When more than one embryo develops within a single seed is called — **Polyembryony**

☆ When seed germinate in the fruit while still attached to the plant is called — **Vivipary**

☆ Embryo development from other than egg cell is called — **Apospory**

☆ Isogenic lines are produced by — **Back cross**

☆ Important male sterility source of sorghum — **Tift 60**

☆ Jerking and tip sterility related to — **Bajra**

☆ Seed certification involve — **5 Phase**

☆ Harvest index is minimum in — **Pulse**

☆ Harvest index is maximum in — **Carrot**

☆ Amino acid deficit in cereals — **Lysine**

☆ Amino acid deficit in pulses — **Methionine**

☆ Harvest index given by — **Donald**

☆ Harvest index also called — **Coefficient of effectiveness**

14
Plant Biotechnology

☆ Biotechnology term was coined by — **Karl Ereky (1919)**

☆ DNA was first synthesized by — **Kornberg (1953)**

☆ RNA was first synthesized by — **Ochoa (1969)**

☆ Father of genetic engineering — **Paul Berg**

☆ PCR was developed by — **Kary Mullis (1980)**

☆ Cry gene was first isolated by — **Baltimore**

☆ First time gene was divided into cistron, muton, recon by — **Benzer (1955)**

☆ Research in the area of tissue culture was first started in India at — **Department of Botany, Delhi University (1960)**

☆ Lal Bahadur Shastri Bio-technology Centre is situated at — **IARI (New Delhi)**

☆ International Centre for Genetic Engineering and Biotechnology (ICGEB) is Situated at — **New Delhi**

☆ Head Quarter of ICGEB at — **Italy**

☆ Applied use of molecular biology and recombinant DNA technology known as — **Biotechnology**

☆ Multiplication of cell of large number of plant placed in appropriate environment conditions with required nutrients is known as — **Plant tissue or *in vitro* culture**

☆ Capability of an isolated single cell to multiply and differentiate into multicellular organism is known as — **Totipotency**

☆ Crossing of two genotypically different plant is — **Hybridization**

☆ Process by which a cell monolayer or a plant explant is transferred without sub division into a fresh medium is called — **Reculture**

☆ *In Vitro* multiplication of plants from a small tissue explant is — **Micropropagation**

☆ Extrachromosomal genetic element found within bacterial cells and replicates independently of chromosomal DNA is — **Plasmid**

☆ Biochemical process or reaction taking place in a test tube — ***In vitro***

☆ Biological process or reaction taking place in a living organism — ***In vivo***

☆ Plant part excised for the *in vitro* cultivation is — **Explant**

☆ Plant part, which is used for regeneration is — **Explant**

☆ Mass of regenerated cells in culture medium is — **Callus**

☆ Alternate forms of gene — **Allele**

☆ Exact genetic replica of a specific gene or an entire organism — **Clone**

☆ Clone are generally — **Heterozygous**

☆ Isolating a gene and process of producing identical copies is — **Gene cloning**

☆ General used nutrient medium in tissue culture — **MS medium and B-5 medium**

☆ Culture of isolated mature or immature embryos — **Meristem culture**

☆ Suspension of free cells of callus in a liqid medium is known as — **Suspension culture**

☆ Young embryo is removed from developing seeds and planted on a suitable nutrient medium *in vitro* is called as — **Embryo culture**

☆ Anther or pollen culture technique is used to obtained — **Haploid plants**

☆ Regeneration of whole plant from anther is called — **Anther culture**

☆ Culture of an organ *in vitro* — **Organ culture**

☆ Method for transforming DNA especially useful of plant cells — **Electroporation**

☆ DNA sequence that codes for a specific polypeptide — **Cistron**

☆ Library composed of complementary copies of cellular mRNA   – **cDNA**

☆ Molecule which encodes genetic information   – **DNA**

☆ Molecules which helps in decoding genetic information carried
   by DNA   – **RNA**

☆ Process of formation of somatic embryo from callus   – **Embryogenesis**

☆ Process of differentiation of shoot and root from the
   somatic embryos is called   – **Organogenesis**

☆ Crossing of plants through fusion of somatic cell   – **Somatic hybridization**

☆ Segment of DNA that codes for a specific characters   – **Gene**

☆ DNA element which has the ability to move from
   one chromosomal position to another   – **Jumping gene**

☆ Bacterium used in genetic engineering   – *E. coli*
                                                  (***Agrobacterium rhizogenes***)

☆ Gene responsible for higher amount of lysine in maize   – **Opaque-2**

☆ TGMS and PGMS systems used for the production of   – **Hybrids**

☆ Hybrid produced using nucleous of one parent cell
   and cytoplasm of both the cell   – **Cybrid**

☆ Process by which genetic material carried by a species
   cell is altered by incorporation of exogenous DNA in
   genome is   – **Transformation**

☆ Serially aligned beades or granules of eukaryotic
   chromosome from local coiling of continuous
   DNA thread is   – **Chromomere**

☆ Triplet of nucleotides in transfer of RNA is   – **Anticodon**

☆ Molecular scissors used in genetic engineering   – **Restriction
                                                                   endonuclease**

☆ Map of genome showing relative position of gene
   and or markers on chromosomes   – **Genetic map**

☆ Complete set of chromosome found in the gemete of
   true diploid is   – **Genome**

☆ Individual with two set of chromosomes are   – **True diploid**

☆ Agar agar was developed by   – **Hesse**

☆ Genetic background of a species is   – **Germplasm**

☆ Cytoplasmic male sterility is used for hybrid seed production of   – **Onion**

☆ Chemical used for induction of Cytoplasmic male sterility
   in barley   – **Silver nitrate**

- ✩ Enzyme that closes nicks or discontinuities in one strand of double stranded DNA by creating a bond is **– DNA ligase**
- ✩ Enzyme that attack bond in DNA **– DNAases**
- ✩ DNA synthesizing enzyme required specifically for replication – **DNA replicase**
- ✩ Single DNA molecule condensed into a compact structure *in vitro* by complexing with accessory histones **– Chromosome**
- ✩ Process of synthesizing multiple copies of a particular DNA sequence **– Gene cloning**
- ✩ Technique of combining 2 or more major genes is **– Gene pyramiding**
- ✩ Process of producing a protein from its DNA and mRNA coding sequences **– Gene expression**
- ✩ In cotton, gene for Helicoverpa resistance has been transferred from soil bacterium **– *Bacillus thruingiensis***
- ✩ Transgenic plants are **– Genetically modified organism**
- ✩ Genotype developed by the process of genetic engineering is called **– Transgenic**
- ✩ DNA amplification is done in the machine **– Thermocycler**
- ✩ Francis Crick`s seminal concept that in nature genetic information generally flows from DNA to RNA to protein **– Central Dogma**
- ✩ One of a number of different forms of a gene is called **– Allele**
- ✩ Group of gene or cloned DNA fragments is **– Gene bank**
- ✩ Sequence of bases at a unique physical location in genome is **– Marker**
- ✩ DNA fragment used as finger print in identification of any organism is **– Molecular markers**
- ✩ PCR technique use for **– Bacterial enzymes for *in vitro* amplification of DNA**
- ✩ PCR technique provides **– Several copies of specific DNA sequence**
- ✩ Transplanting a cell, tissue or organ from one nutrient medium to another **– Subculture**
- ✩ First biotech plant is **– Tobacco**
- ✩ First sequenced plant is **– *Arabidopsis thaliana* (weed)**
- ✩ Bt transgenic cotton variety **– Coker-312**
- ✩ Bt transgenic cotton hybrids developed by **– MAHYCO**

☆ Bt genes are introduced in cotton against the pest — **Cotton Boll worm**

☆ Vegetable crop under approval for Bt technology — **Brinjal**

☆ Terminator technology is recently used in — **Cotton**

☆ β-carotene is — **Precursor of Vitamin A**

☆ Golden rice provides — **Higher β-carotene over normal rice**

☆ Project on Indian mustard oil with higher β-carotene is initiated by — **TERI**

☆ Commonly used molecular marker or DNA markers are — **RFLP, RAPD, CAPS, SSR, DNA finger printing**

☆ 14C has a half life of — **5730 years**

☆ Micro sporogenesis occur in — **Stamens**

☆ Marker Aided selection (MAS) is also known as — **Marker Assisted Selection**

☆ ELISA denotes — **Enzyme Linked Immunosorbent Assay**

☆ PCR denotes — **Polymerase Chain Reaction**

☆ RFLP denotes — **Restriction Fragment Length Polymorphism**

☆ RAPD denotes — **Random Amplified Polymorphic DNA**

☆ STMS denotes — **Sequence Tagged Micro satellite Site**

☆ SNP denotes — **Single Nucleotide Polymorphism**

☆ TERI denotes — **Tata Energy Research Institute**

15
Entomology

Paddy

- ☆ Scientific name of yellow stem borer — *Scirpophaga incertulas*
- ☆ Paddy stem borer is — **Monophagous**
- ☆ Family of stem borer — **Pyraustinae**
- ☆ Scientific name of paddy gall midge — *Orseolia oryzae*
- ☆ Family of gall midge — **Cecidomyidae**
- ☆ Shiver shoot or onion leaf is caused by — **Gall midge**
- ☆ Scientific name of green leaf hopper (GLH) — *Nephotettix nigropictus*
- ☆ Family of green leaf hopper — **Cicadellidae**
- ☆ Vector of paddy tungro is — **Green leaf hopper (GLH)**
- ☆ Scientific name of white backed plant hopper (WBPH) — *Sogatella furcifera*
- ☆ Scientific name of brown plant hopper (BPH) — *Nilaparvata lugens*
- ☆ Family of brown plant hopper — **Delphacidae**
- ☆ Vector of grassy stunt disease is — **Brown plant hopper (BPH)**
- ☆ Scientific name paddy case worm — *Nymphula depunctalis*
- ☆ Scientific name of Paddy army warm — *Mythimna separata*
- ☆ Scientific name of gundhi bug — *Leptocorisa acuta*

☆ Family of gundhi bug — **Alydidae**

☆ Causing stage of Gundhi bug is — **Milking stage**

☆ Chaffy grains with black spot is the infestation of — **Gundhi bug**

☆ Gundhi bug emit a — **Repugnant smell**

☆ Scientific name of leaf folder — *Cnapholocrocis medinalis*

☆ Family of leaf folder — **Pyraustidae**

☆ Major pest of paddy — **Yellow stem borer**

☆ Clipping off the top of paddy seedlings containing immature stages of insects reduces the carry over of infestation of — **Paddy hispa**

☆ Which pest is responsible for Severe damage to paddy panicle at night — **Paddy army worm**

☆ Important nematode pest on paddy — **Aphelenchoides**

☆ *Trichogramna* is a — **Egg parasitoid**

☆ Ufra disease of paddy is due to — *Ditylenchus angustus* **(Nematode)**

Soybean

☆ Scientific name of girdle beetle — *Oberea brevis*

☆ Scientific name of stem fly — *Melanagromyza sojae*

☆ Which pest is known as stem borer of soybean — **Girdle beetle**

Groundnut

☆ Scientific name of groundnut aphid — *Aphis craccivora*

☆ Scientific name of groundnut leaf miner — *Aproaerema modicella*

☆ Scientific name of white grub — *Phyllophaga errors*

Wheat

☆ Scientific name of pink stem borer — *Sesamia inferens*

☆ Scientific name of termite — *Odontotermes obesus*

☆ Scientific name of black cut worm — *Agrotis ipsilon*

☆ Ear cockle nematode is — *Anguina tritici*

☆ Tundu/yellow ear rot disease due to — *Anguina tritici* + *Carynebacterium tritici*

☆ Control of tundu disease — **Hot water treatment 50°C for 2 hrs**

☆ Pink stem borer attacks to plant in — **Night**

☆ Which pest attacks all the parts of the plant   **– Termite**

☆ Serious pest of wheat   **– Termite and White grub**

☆ White grub is   **– Phyllophaga**

Mustard

☆ Scientific name of aphid   *– Lipaphis erysimi*

☆ Scientific name of mustard saw fly   *– Athalia proxima*

☆ Order of mustard saw fly   **– Hymenoptera**

☆ Scientific name of painted bug   *– Bargrada cruciferarum*

Pigeon Pea

☆ Scientific name of pod borer   *– Etiella zinckenella*

☆ Scientific name of plume moth   *– Exelastis atomosa*

☆ Scientific name of pod fly   *– Melanagromyza obtusa*

☆ Scientific name of pod bug   *– Clavigralla gibbosa*

Chick Pea

☆ Scientific name of pod borer   *– Helicoverpa armigera*

☆ Scientific name of cut worm   *– Agrotis ipsilon*

☆ Major pest of chick pea is   **– Gram pod borer and cut worm**

☆ Greasy cut worm attack in   **– Night**

Cotton

☆ Scientific name of pink boll worm   *– Pectinophora gossypiella*

☆ Scientific name of spotted boll worm   *– Earias vitella*

☆ Scientific name of American boll worm   *– Helicoverpa armigera*

☆ Scientific name of white fly   *– Bemisia tabaci*

☆ Scientific name of red cotton bug   *– Dysdercus koenigii*

☆ Scientific name of cotton jasid   *– Amrasca bigutulla*

☆ Red cotton bug lay its egg   **– On Soil**

☆ Pest causing flaring of squares in cotton   **– Spotted boll worm**

☆ Vector of leaf curl virus   **– White fly**

☆ Family of white fly   **– Aleyrodidac**

☆ Bt formulation is used for   **– Early instars of bollworms**

☆ Which pest causes hopper burn — **Cotton jassid**

☆ Boll worm is a dangerous insect of — **Cotton**

☆ Highest consumption of pesticide found in — **Cotton**

☆ Main symptom of American wall worm — **Larger circular bore holes with faecal pellets**

☆ Rosetting of flower and double seed formation in the symptoms of — **Pink boll worm**

☆ Curling of leaf upwards and yellowing of terminal cotton shoots is a characteristics symptom of presence of — **Cotton aphid**

Sugarcane

☆ Scientific name of leaf hopper/pyrilla — *Pyrilla purpusilla*

☆ Scientific name of top borer — *Tryporyza novella*

☆ Family of top borer — *Pryalidae*

☆ Dead heart in sugar cane is caused by — **Shoot borer**

☆ Scientific name of sugar cane shoot borer — *Chilo infuscatellus*

☆ Family of shoot borer — **Crambidae**

☆ Scientific name of sugar cane root borer — *Emmalocera depressella*

☆ Scientific name of sugarcane white fly — *Aleurolobus barodensis*

☆ Which pest causes bunchi top appearance in sugarcane — **Top borer**

☆ Greasy cut worm attack in — **Night**

☆ Wooly aphid is the serious pest of — **Sugarcane**

☆ Biological control of borer — *Trichogramma japonicum*

Potato

☆ Scientific name of aphid — *Aphis gossypii*

☆ Scientific name of potato tuber moth — *Phthorimaea operculella*

Mango

☆ Scientific name of mango hopper — *Amritodus atkinsoni*

☆ Scientific name of mango mealy bug — *Drosicha mangiferae*

☆ Sticky bands around tree trunks provide protection against — **Mango mealy bug**

☆ Major pest of mango — **Mealy bug**

Citus

☆ Scientific name of fruit sucking moth of citrus — *Otheris materna*

☆ Scientific name of lemon butter fly — *Papillio demoleus*

☆ Scientific name of citrus psylla — *Diaphorina citri*

☆ Major pest of citrus — **Psylla**

Pomegranate

☆ Scientific name of fruit borer of pomegranate — *Conogethes punctiferalis*

☆ Major pest of pomegranate — *Virachola isocrates*

Apple

☆ Scientific name of woolf aphis of apple — *Eriosoma lanigerum*

☆ Major pest of apple — **San jose scale**

Important Pest of other different Crops

☆ Scientific name of papaya fruit fly — *Bactrocera dorsalis*

☆ Scientific name of banana stem weevil — *Odoiporus longicollis*

☆ Scientific name of guava fruit fly — *Bactrocera diversus*

☆ Scientific name of tobacco cut worm — *Spodoptera litura*

☆ Scientific name of sorghum shoofly — *Atherigona varia soccata*

☆ Scientific name of sunflower head borer — *Helicoverpa armigera*

☆ Scientific name of tomato fruit borer — *Helicoverpa armigera*

☆ Scientific name of brinjal fruit and shoot borer — *Leucinodes orbonalis*

☆ Major pest of brinjal — **Hadda beetles**

☆ Scientific name of fruit and shoot borer of okra — *Earias vitella*

☆ Scientific name of chilli thrips — *Thrips tabaci*

☆ Scientific name of red pumpkin beetles of cucurbits — *Raphidopalpa foveicollis*

☆ Scientific name of fruit fly of cucurbits — *Dacus cucurbitae*

☆ Scientific name of diamond back moth of cabbage — *Plutella xylostella*

☆ Scientific name of cabbage head borer — *Hellula undalis*

☆ Major pest of cabbage — **Diamond back moth**

Store Grain Pest

- ☆ Scientific name of khapra beetle or wheat beetle — *Trogoderma granarium*
- ☆ Scientific name of red flour beetle — *Tribolium castaneum*
- ☆ Scientific name of pulse beetle — *Callosobruchus chinensis*
- ☆ Scientific name of rice moth — *Corcyra cephalonica*

Important Points

- ☆ Insecticide act was passed — **1968**
- ☆ Destructive insect and pest act was passed — **1914**
- ☆ Plant protection and Quarantine act was passed — **1912**
- ☆ IPM term was given by — **Geiger and Clark (1961)**
- ☆ First pesticide was discovered by — **Millardet (1882)**
- ☆ First insecticide discovered was — **DDT**
- ☆ DDT was discovered by — **Zeidler**
- ☆ Central plant protection training institute (NPPTI) located at — **Hyderabad**
- ☆ Safest insecticide for honey bee is — **Endosulphon**
- ☆ Pest occurs most frequently on cultivated crop is called — **Regular pest**
- ☆ Pest occurs throughout the year on crops is called — **Persistent pest**
- ☆ Pest occurs in few isolated localities is called — **Sporadic pest**
- ☆ Pest occurs in same area of year after year is called — **Endemic pest**
- ☆ Pest occurs in area in severe form is called — **Epidemic pest**
- ☆ Pest occurs in a large geographical area or entire country is called — **Pandemic pest**
- ☆ Pest population should be kept below — **Economic threshold level (ETL)**
- ☆ Sequence of seed treatment is — **Fungicide→ Insecticide→ Miticide →Rhizobium**
- ☆ Blood colour of insects — **Green with yellow**
- ☆ Entomology word derived from — **Greek**
- ☆ Bacillus thuringiensis, NPV are — **Biopesticide**
- ☆ Crop showing maximum resistance to nematode is — **Marigold**
- ☆ Lac research station is situated at — **Bihar**
- ☆ Scientific name of Lac — *Laccifer lacca*
- ☆ Order of lac — **Hemiptera**

☆ Silvery shoot of onion is caused by — **Thrips**

☆ Rat poison known as — **Zinc phosphide (Zn_3P_2)**

☆ Full form of DDT — **Dichloro Diphenyl Trichloroethane**

☆ Full form of BHC — **Benzene Hexa Chloride**

☆ Trade name of BHC is — **Gammexane**

☆ MCPA is a — **Herbicide**

☆ International pest is — *Schistocerca gregaria*

☆ National plant protection training institute (NPPTI) was established in — **1966**

☆ Pheromone traps attracts — **Male moths**

☆ First plant parasitic nematode is — *Anguina tritici*

☆ Which state is the largest producer of lac in India — **Bihar**

☆ Lady bird beetle is a — **Predator of aphid**

☆ Which pesticide is suitable for the control of white grub — **Phorate**

☆ Malaria was first identified in which place in India — **Hyderabad**

☆ Root knot resistant variety of tomato — **Pusa ruby**

☆ Most common example of mechanical control of insect is use of — **Trap cropping**

☆ Setaceous antennae is present in — **Cockroach**

☆ Hindgut of an insect in also known as — **Proctodeum**

☆ Antennae are absent in — **Protura**

☆ Chitin is absent in — **Epicuticle**

☆ Dengu is transmitted by vector — *Aedes aegypti*

☆ Fish meal trap used which pest — **Sorghum shoot fly**

☆ First insect growth regulator is — **Methoprene**

☆ Air tight container are used against the storage pest — **Khapra beetle**

☆ Largest phylum in animal kingdom — **Arthropoda**

☆ Largest order in class insecta is — **Coleptera**

☆ Safe insecticide for vegetable is — **Malathion**

☆ True fly is also known as — **House fly**

☆ Blood sugar in insect is — **Trehalose**

☆ Halters are present in — **Diptera order insect**

- ☆ Neem seed kernel extract (NSKE) contains — **Hexanortriterpenoids**
- ☆ Dwelling place of termites is also known as — **Mound**
- ☆ Insect procuticle is made up of — **Two layer**
- ☆ Important nematode on vegetables is — ***Pratylenchus***
- ☆ NPV is very effective in the control of — ***Helicoverpa armigera***
- ☆ Anemophily — **Pollination by wind**
- ☆ Hydrophily — **Pollination by water**
- ☆ Ornithophily — **Pollination by bird**
- ☆ Entomophily — **Pollination by insects**
- ☆ Antennae are absent in — **Proturans**
- ☆ First fused abdominal segment of hymenoptera is known as — **Propodeum**
- ☆ Carmine red dye is extracted from — **Cochineal insects**
- ☆ Forewings of stylops are called — **Halteres**
- ☆ *Aedes aegypti* is a vector of — **Dengue**
- ☆ Diamond back moth is major pest of — **Crucifers**
- ☆ Asymmetrical mouth parts are found in — **Thrips**
- ☆ Nymph of dragon fly is known as — **Naid**
- ☆ Outer wall of egg is known as — **Chorion**
- ☆ Haemoglobin is found in — **Chironomid larva**
- ☆ Honey dew is secreted by — **Aphid**
- ☆ Peritrophic membrane is found in — **Mid gut**
- ☆ Insect thorax consists of — **3 segments**
- ☆ Extremely mammalian toxic pesticides label is — **Bright Red**
- ☆ Highly toxic pesticides label is — **Dark yellow**
- ☆ Number of chambers in heart of cockroach is — **13**
- ☆ Summer diapause is known as — **Aestivation**
- ☆ Major component of insect cuticle is — **Chitin**
- ☆ Locust, Termite, Bihar hairy caterpillar, White grub are example of — **Polyphagus pest**

16
Apiculture

☆ Apiculture word derived from	**– Latin word**
☆ Apiculture is also known as	**– Bee keeping**
☆ Location where bee are kept is called	**– Apiary**
☆ Bee keeper is also known as	**– Apiarist**
☆ Scientific name of Rock bee	*– Apis dorsata*
☆ Scientific name of Little bee	*– Apis florae*
☆ Scientific name of Indian bee	*– Apis indica*
☆ Scientific name of European bee	*– Apis mellifera*
☆ Scientific name of Stingless bee	*– Trigona iridiopennis*
☆ Family of honeybee	**– Apidae**
☆ Honey bee belongs to order of	**– Hymenoptera**
☆ Stingless bee is found in	**– Kerala**
☆ Good honey gather species of honey bee	*– Apis dorsata*
☆ Yield of *Apis dorsata*	**– 50-80 kg/colony/year**
☆ Yield of Trigona iridiopennis	**– 300-400 g/Year**
☆ Thaisac brood disease is caused by	**– Virus**
☆ Stingless bee found in	**– Kerala**
☆ Central honey bee research station is situated at	**– Pune**

- ☆ Sting in worker bee is a modified part of — **Ovipositor**
- ☆ Male of honey bee is called — **Drone**
- ☆ Honey bee pollen basket is present in — **Hind leg**
- ☆ Functional female of the bee colony is — **Queen**
- ☆ Largest size of honey bee is — *Apis dorsata*

17
Sericulture

- ☆ Scientific name of silk worm — *Bombyx mori*
- ☆ Family — **Bombycidae**
- ☆ Scientific name of muga silk — *Antheraea assamensis*
- ☆ Rearing of silkworms for the production of raw silk is known as — **Sericulture**
- ☆ Growing mulberry for rearing of silk worm is called — **Moriculture**
- ☆ Sericulture is also known as — **Silk farming**
- ☆ Silk is a continuous filament fibre consisting of — **Fibroin protein**
- ☆ Food of Silk worm larvae is — **Mulberry leaves**
- ☆ How much silkworms are required to produce 1 kg of silk — **5500**
- ☆ Main host tree of silkworm is — **Mulberry**
- ☆ Central Sericulture Research and Training Institute (CSRTI) situated at — **Mysore (Karnataka)**
- ☆ Central Silk Research Station situated at — **West Bengal**
- ☆ Pebrine disease is found in — **Silk worm**
- ☆ Life cycle of mulberry silk worm is completed in about — **40-45 Days**
- ☆ Silk gland in silkworm are modified — **Labial gland**
- ☆ Stifling operation in silkworm is done for — **Killing the pupae**

18
Pathology

Paddy

- ☆ Causal organism of brown spot of paddy — *Helminthosporium oryzae*
- ☆ Causal organism of bacterial leaf spot — *Xanthomonas oryzae*
- ☆ Causal organism of paddy blast — *Pyricularia oryzae*
- ☆ Causal organism of sheath blight of paddy — *Rhizoctonia solani*
- ☆ Poor man's disease of paddy — **Bacterial leaf blight**
- ☆ Most destructive phase of the bacterial blight of paddy is — **Kresek**
- ☆ Brown spot of paddy is — **Externally seed borne disease**
- ☆ Blast of paddy is — **Air borne disease**
- ☆ Sheath blight of paddy is — **Soil borne disease**
- ☆ Paddy blast is effectively controlled by spraying of — **Edifenphos**
- ☆ Khaira disease of paddy is caused by — **Zinc deficiency**
- ☆ Khaira disease of paddy is controlled by spraying of — **Zinc sulphate (5 kg) + Lime (2.5 kg) per ha in 10 days nursery**
- ☆ Main symptom of tungro disease of paddy — **Yellowing of leaves**
- ☆ Vector of tungro disease — **GLH**

☆ Disease responsible for the great Bengal famine in 1942-43 — **Brown leaf spot of paddy**

☆ Montek disease of paddy is caused by — **Paddy root nematode**

☆ Bakanae disease of paddy also called — **Foot rot**

☆ Causal organism of foot rot — *Gibberella fujikuroi*

Ground Nut

☆ Causal organism of early leaf spot — *Cercospora arachidicola*

☆ Causal organism of late leaf spot — *Cercospora parsonata*

☆ Tikka disease of groundnut is also called — **Leaf spot**

☆ Vector for bud necrosis of groundnut is — **Thrips**

☆ Causal organism of collar rot — *Aspergillus niger*

☆ Causal organism of rust — *Puccinia arachidis*

Soybean

☆ Causal organism of yellow mosaic — **Mungbean yellow mosaic virus (MYMV)**

☆ Causal organism of anthracnose/pod blight — *Colletotrichum truncatum*

Cotton

☆ Causal organism of wilt — *Fusarium oxysporum*

☆ Causal organism of black arm or bacterial blight — *Xanthomonas compestris*

☆ Causal organism of bacterial blight/ angular leaf spot — *Xanthomonas malvacearum*

☆ Causal organism of Root rot — *Rhizoctonia bataticola*

☆ Causal organism of anthracnose — *Colletotrichum capsici*

☆ Causal organism of leaf blight — *Alternaria macrospora*

☆ Bacterial blight of cotton is — **Internally seed borne**

☆ Wilt of cotton is — **Seed and soil borne**

Sugarcane

☆ Causal organism of red rot — *Colletrotricum falcatum*

☆ Causal organism of red stripe — *Phytomonas rubilineans*

☆ Causal organism of smut — *Ustilago scitaminea*

☆ Causal organism of whip smut   – *Ustilago hordei*

☆ Most serious disease of sugarcane   – **Red rot**

☆ Grassy shoot of sugarcane caused by   – **MLO**

☆ Pith of the red rot affected can emits   – **Rotten fish like smell**

Wheat

☆ Causal organism of karnal bunt   – *Neovossia indica*

☆ Causal organism of loose smut   – *Ustilago nuda tritici*

☆ Causal organism of black/stem rust   – *Puccinia graminis tritici*

☆ Causal organism of yellow/strip rust   – *Puccinia graminis striformis*

☆ Causal organism of brown/orange/leaf rust – *Puccinia graminis recondita*

☆ Molya disease is due to   – *Heterodera avcnae*

☆ Which disease of wheat is discovered in Haryana   – **Karnal bunt**

☆ Karnal bunt of wheat is   – **Soil, air, seed borne disease**

☆ Foul smell of karnal bunt infection field is due to   – **Trimethyl-amine**

☆ Loose smut of wheat is   – **Internally seed borne disease**

☆ Loose smut of wheat can be controlled by seed treatment of   – **Vitavax**

☆ Solar heat treatment is used to control   – **Loose smut of wheat**

☆ Which rust was earliest appeared in India on wheat   – **Brown/orange/ leaf rust**

☆ Alternate host of black stem rust of wheat in India is   – *Berberis* **spp.**

☆ Which stage of wheat rust fungus is considered as the perfect stage   – **Telial stage**

☆ Bacterial rot of wheat ears is also known as   – **Spike blight/tundu/ yellow slime disease**

Barley

☆ Causal organism of covered smut of barley   – *Ustilago hordei*

☆ Sooty or charcoal like powdery mass usually appearing on floral organs particularly the ovary is   – **Smut**

Maize

☆ White bud of maize is caused by   – **Zinc deficiency**

Mustard

- ☆ Causal organism of alternaria blight — *Alternaria brassicae*
- ☆ Causal organism of Blister/white rust — *Albugo candida*
- ☆ White rust of crusifers is — **Pseudo rust**

Sunflower

- ☆ Causal organism of alternaria blight or leaf spot — *Alternaria helianthi*
- ☆ Causal organism of root and collar rot — *Sclerotium rolfssi*

Sorghum

- ☆ Causal organism of grain smut — *Sphacelotheca sorghi*
- ☆ Causal organism of head smut — *Sphacelotheca relliana*
- ☆ Grain smut of sorghum is also known as — **Covered/kernel/shoot smut**
- ☆ Grain and head smuts of sorghum is — **Seed borne disease**
- ☆ Most serious smut among the smuts affecting sorghum is — **Grain smut**

Bajra

- ☆ Causal organism of ergot disease of bajra — *Claviceps fusiformis*
- ☆ Causal organism of green ear disease/ downey mildew of bajra — *Sclerospora graminicola*
- ☆ Ear showing honey dew symptom is characteristics feature of — **Grain smut of bajra**
- ☆ Smut disease infect the plant at — **Tillering stage**

Pigeon Pea/Arhar

- ☆ Causal organism of wilt of pigeon pea — *Fusarium oxysporium* spp. *udum*
- ☆ Vector of sterility mosaic of pigeon pea — **Mite**
- ☆ Scientific name of mite — *Aceria cajani*

Brinjal

- ☆ Causal organism of wilt in brinjal — *Pseudomonas solanacearum*
- ☆ Causal organism of phomopsis blight/fruit rot — *Phomopsis vexans*
- ☆ Damping off of brinjal seedlings is due to — *Pythium* spp.

Potato

- ☆ Causal organism of late blight — *Phytophthora infestans*
- ☆ Causal organism of early blight — *Alternaria solani*

☆ Causal organism of wart disease — *Synchytrium endobioticum*

☆ Causal organism of black scurf — *Rhizoctonia solani*

☆ Most dangerous disease of potato — **Late blight**

☆ Tuber borne disease of potato is — **Black scurf**

☆ Which potato disease caused Irish famine in 1845 — **Late blight of potato**

☆ Potato virus disease are spread by — **Aphids**

☆ Black heart of potato caused by — **Lack of Oxygen**

Tomato

☆ Causal organism of early blight — *Alternaria solani*

☆ Leaf curl of tomato is spread by — **White fly**

Tobacco

☆ Causal organism of damping off — *Pythium aphanidermatum*

☆ Tobacco mosaic disease is caused by — **Nicotiana virus-1**

☆ Root knot disease of tobacco is effective controlled by — **Carbofuran**

☆ Orobanche parasite is associated with — **Tobacco**

Cabbage

☆ Causal organism of white blister — *Albugo candida*

☆ Causal organism of black rot — *Xanthomonas compestris*

☆ Which disease is more severe in acidic soil is — **Black rot**

Mango

☆ Causal organism of mango malformation — *Fusarium moniliforme*

☆ Mango necrosis/black tip is caused by — **Boron deficiency**

☆ Mango malformation is common in — **North-West India**

Citrus

☆ Causal organism of citrus canker — *Xanthomonas compestris*

☆ Causal organism of citrus gummosis — *Phytophthora palmivora*

☆ Mottle leaf of citrus is due to — **Zinc deficiency**

☆ Greening of citrus is caused by vector transmitted Pathogen.
The causative agents are motile bacteria — **Candidatus Liberibacter spp.**

Banana

☆ Panama wilt is also called — **Fusarium wilt**

☆ Bunchy top of banana is caused by — **Virus**

☆ Panama wilt of banana is — **Fungal disease**

☆ Moko disease of banana is — **Bacterial disease**

Important Disease of other Crops

☆ Causal organism of guava anthracnose — *Collectotrichum psidii*

☆ Causal organism of apple scab disease — *Venturia inaequalis*

☆ Leaf curl and mosaic of papaya is caused by — **Virus**

☆ Causal organism of cauliflower club root — *Plasmodiophora brassicae*

☆ YVMV of okra is transmitted by — **White fly**

☆ Causal organism of sesame phyllody — **MLO (Mycoplasma)**

☆ Causal organism of wilt of chick pea — *Fusarium oxysporium* **spp.** *ciceri*

☆ Causal organism of chilli anthracnose/ ripe rot/die back — *Collectotrichum capsici*

Important Points

☆ Father of pathology — **Anton de Bary**

☆ Father of Modern plant pathology — **E. J. Butler**

☆ Father of microbiology — **Louis Pasteur**

☆ Father of Indian pathology — **E. J. Butler**

☆ Father of mycology — **P. A. Micheli**

☆ Father of nematology — **Bastian**

☆ Father of plant virology — **Beijerinck**

☆ Father of plant bacteriology — **E. F. Smith**

☆ Bordeaux mixture was developed by — **P.M.A. Millardet**

☆ Life cycle of wheat rust given by — **K. C. Mehta**

☆ Bacterium was discovered by — **Leeuwenhoek**

☆ First plant protection advisor to government of India — **H. S. Pruthi**

☆ Which disease was discovered in Haryana — **Karnal bunt**

☆ Crystallization of virus by — **Stanley (1935)**

☆ Electron microscope discovered by — **Knoll and Ruska (1932)**

☆ Ray fungi is the name given to — **Actinomycetes**

☆ Bacterial are commonly seen in — **Neutral soil**

☆ Milky disease is caused by — *Bacillus popilliae*

☆ Alginates are isolated from — **Phaeophyta**

☆ Bacillus is a — **Thermophilic bacteria**

☆ Sterols are present in — **Mycoplasma**

☆ Which crop showing maximum resistance to nematode — **Marigold**

☆ Which fungicide use for control of smut — **Vitavax**

☆ Bordeaux mixture used as — **Fungicide**

☆ Proportion of contents in Bordeaux mixture — **Copper sulphate : Lime : Water (5:5:50)**

☆ Major disease in rose — **Dieback**

☆ Damping off in seedlings stage is due to — *Pythium* **spp.**

☆ Causal organism of powdery mildew in cucurbits — **Erysiphe cichoracearum**

☆ Causal organism of downey mildew in cucurbits — *Pseudoperonospora cubensis*

☆ Downey mildew disease can be effectively managed by spraying of — **Metalaxyl**

☆ Mycorrhiza is a symbiotic association between — **Fungi and Root of higher plant**

☆ Which pathogen caused heavy losses to wine industry in France due to its epidemics in 1875 — *Plasmosphora viticola*

☆ Pathogen associated with the discovery of Bordeaux mixture is — *Plasmosphora viticola*

☆ Downy mildew of grape wine is controlled by — **Bordeaux mixture**

☆ Sporulation in bacteria occurs at — **Lag phase**

☆ Biggest virus — *Citrus tristeza*

☆ Vitavax is — **Systemic fungicide**

☆ Total root parasite is — **Orobanche**

☆ Semi root parasite is — **Striga**

☆ Total stem parasite is — **Cascuta**

☆ Semi stem parasite is — **Loranthus**

☆ Plant pathogen made up of only SS RNA no protein is there — **Viroid**

☆ Plant virus mostly having — *ss* **RNA**

☆ Tobacco mosaic virus — *ss* **RNA**

☆ Gemini virus — *ss* DNA

☆ Reo virus — *ds* RNA

☆ Caulimo virus, Bacteriophage — *ds* DNA

☆ Mancozeb, Zenab, Thiram are — **Dithiocarbamates**

☆ Endosulphan is a — **Chlorinated hydrocarbons**

☆ Agrosan, Ceresan and Aretan are — **Organomercurial fungicide**

☆ Aldicarb, Carbaryl are — **Carbamates and thio salts**

☆ Infectious protein molecule called — **Prions**

☆ First discovered viroid disease is — **Spindle tuber**

☆ Spindle tuber disease occurs in — **Potato**

☆ Powdery mildew control by — **Sulphur fungicide**

☆ Mycoplasma is larger than virus but smaller than — **Bacteria**

☆ Nucleic acid + protein — **Virus**

☆ RNA + Protein — **Plant virus**

☆ DNA + Protein — **Animal virus**

☆ Plant viroid — **RNA only**

☆ Viroid — **Nucleic acid only**

☆ Tail of bacteriophage is composed from — **Protein only**

☆ Virus particle called as — **Virions**

☆ Pathogen of xanthomonas comes in environment in the form of — **Ooze**

☆ Ooze test is done for detecting of — **Bacteria**

☆ Mushroom is — **Basidiocarp**

☆ Thermophilic bacteria generally grown in the temperature of — **50-70°C**

☆ Which crop is known as vegetarian meat — **Mushroom**

☆ Break down of protectin is saprophytically by bacteria, it is also known as — **Putrefication**

☆ In which bacterial reproduction phase, virus is involved — **Transduction**

☆ Aplanospores are produced by — **Fungi and Algae**

☆ Katte disease is also known as — **Marble disease**

☆ Tyndallization is performed at — **100°C for 30 min**

☆ Powdery mildew is favoured by — **Warm days with cool night**

- ☆ How much time is needed for sterilization in autoclave — **–15 Minutes**
- ☆ Enzyme used in ELISA for detecting virus — **– Alkaline phosphatase**
- ☆ Largest bacteria — **– *Beggiatoa mirabilis***
- ☆ Smallest bacteria — **– Dialister pneumosintes**
- ☆ Inter croppig of potato + garlic is found useful to control — **– Late blight and tuber rot disease**

19
Mushroom Cultivation

- ☆ National Research Centre for Mushroom (NRCM) — **Solan (Himachal Pradesh)**
- ☆ Kingdom of mushroom is — **Fungi**
- ☆ Mushroom term derived from — **French word**
- ☆ Scientific name of white button mushroom — *Agaricus bisporus*
- ☆ Scientific name of Paddy straw mushroom — *Volvariella Volvacea*
- ☆ Scientific name of Oyster mushroom — *Pleurotus ostreatus*
- ☆ Family of mushroom — **Hypocreaceae**
- ☆ Order of mushroom — **Agaricales**
- ☆ Mushroom is fleshy, spore bearing fruiting body of a — **Fungus**
- ☆ Most common used mushroom is — **White button mushroom**
- ☆ Classification of mushroom is based on the colour of — **Spore print**
- ☆ Which mushroom species is suitable for summer or tropical area — **Paddy straw mushroom**
- ☆ Which mushroom species is requires low temperature — **White button mushroom**
- ☆ Optimum temperature for growing Oyster mushroom is — **22°C to 28°C temperature**
- ☆ Mushroom growers used propagating material for planting beds is called — **Spawn**

20
Agriculture Extension

- ☆ First basic persuasive communication model given by — **Aristotle**
- ☆ The term "Extension Education" was first coined in — **UK**
- ☆ The word " Extension" was used first time in — **USA**
- ☆ Extension activity was started first time in — **USA**
- ☆ The father of extension in India — **K. N. Singh**
- ☆ The father of University extension — **James Stuart**
- ☆ The father of extension in USA — **Dr. Seaman A. Knapp**
- ☆ The father of demonstration in extension — **Dr. Seaman A. Knapp**
- ☆ The father of extension education — **J. P. Leagons**
- ☆ The first use of the word extension was done by — **Voorhees**
- ☆ Linear model of communication was given by — **J. P. Leagons**
- ☆ The concept of feedback was introduced by — **Berlo's**
- ☆ First model of communication was given by — **Berlo's**
- ☆ Communication term is derived from — **Latin word**
- ☆ Broadcasting through radio started in India on — **23 July 1927**
- ☆ Vividh Bharti launched — **1957**
- ☆ The first telecast on television in India was started in — **15 Sept. 1959**
- ☆ Radio an example for — **Hot media**

- ☆ The national press day is observed on — **16 November**
- ☆ Balance theory of communication is proposed by — **Heider**
- ☆ Theory of communication act is proposed — **New comb.**
- ☆ Abbreviation of COIK means — **Clear Only If Known**
- ☆ ABC of good news or journalism — **Accuracy, Brevity, Clarity**
- ☆ ABC of poster — **Attractive, Brevity, Clarity**
- ☆ Optimum number of flash card is — **10-12 cards**
- ☆ Bulletin should contain — **24-48 pages**
- ☆ Black and white colour is — **Achromatic colour**
- ☆ HUE indicates name of — **Colour**
- ☆ Primary colour used in extension is — **Red, Blue, Yellow**
- ☆ Primary colour are — **Red, Blue, Green**
- ☆ Extension term have originated in — **Latin roots**
- ☆ Extension literally means — **Stretching out**
- ☆ "Cone of experience" was devised/given by — **Edger Dale**
- ☆ Diffusion has — **6 elements**
- ☆ Stage of adoption process has — **5**
- ☆ 4H — **Head, Hand, Heart, Health**
- ☆ Kisan call centre toll free number from landline is — **1551**
- ☆ Chronological order of activities is otherwise known as — **Plan of work**
- ☆ A Predetermined course of action is called — **Plan**
- ☆ The term evaluation was derived from — **Latin**
- ☆ The difference between what is and what ought to be is — **Need**
- ☆ Entrepreneur derived from — **French word**
- ☆ The term "Entrepreneur" was coined by — **Richard Cantillon**
- ☆ Five year plan is a — **Medium term plan**
- ☆ The current five year plan is — **12th (2012-2017)**
- ☆ Chairman of NITI Aayog is — **Prime Minister**
- ☆ First chairman of national planning commission of India — **J. L. Nehru**
- ☆ Govt. of India set up planning commission in — **1950**
- ☆ Panchayat Raj was first introduced in — **Nagaur (Rajasthan)**
- ☆ Old name of NITI Aayog — **Planning Commision**

☆ Presently the "Panchayati Raj" in India has — **Three tier system**

☆ General body of panchayat is — **Gram Sabha**

☆ First tier of panchayati raj is — **Gram panchayat**

☆ Panchayati raj came after — **Self determined programme**

☆ First KVK in India was set up under — **TNAU**

☆ Old name of KVK was — **Agriculture Polytechnic**

☆ KVK also called — **Farm Science Center**

☆ Basic element of communication — **3**

☆ IRDP, TRYSEM, DWCRA, SITRA, GKY, MWS has been merge under — **SGSY**

☆ NREGA act developed on — **7 December 2005**

☆ NREGA implemented on — **2 February 2006**

☆ Panchayati raj system recommended by B.R. Mehta in — **2 Oct. 1959**

☆ The constitutional 73rd amendment act was passed in — **1992**

☆ The Head of National Development Council (NDC) is — **Prime Minister**

☆ National Development Council (NDC) was set up in — **1952**

☆ T and V System of extension was first introduced in — **Turkey**

☆ The total number of KVK in India is — **638**

☆ Ideal no. for training is — **25-30 person**

☆ Royal commission on agriculture was established in — **1928**

☆ Father of PRA — **Robert Chamber**

☆ Extension education institutes (EEIs) presently in India is — **4**

☆ Headquarters of CAPART are at — **New Delhi**

☆ Food cooperation of India (FCI) was set up in — **1965**

☆ NABARD act was passed in — **1979**

☆ Lass well's model also called — **Rudimentary model of communication**

☆ Shanon weaver model also called — **Mathematical theory of communication**

☆ Communication is a — **Two way process**

☆ Communication is an organization flows — **Downward, Upward, and Horizontally.**

- ☆ Lab to Land Programme was launched in — **July1979 (Golden Jubilee of ICAR)**
- ☆ T and V system was proposed by — **Daniel Benor and D. Boxster**
- ☆ Panchayat raj system was adopted first in — **Rajasthan and Andhra Pradesh**
- ☆ Minimum number of days of employment a year guaranteed under the MNREGA — **100 days**
- ☆ NABARD came into existence on — **12 July 1982**
- ☆ NAARM is located at — **Hyderabad**
- ☆ "Operation Flood " is related to — **Dairy Development**
- ☆ Father of white revolution in India — **Dr. Verghese Kurien**
- ☆ The outer ceiling for the cost of an individual project in agri –clinics is — **10 lakhs**
- ☆ The e-chaupal concept in extension delivery is promoted by — **ITC**
- ☆ The number of SAUs in India is — **43**
- ☆ The first farm magazine in India is — ***Khethi***
- ☆ The national press day is observed on — **16 November**
- ☆ "Krishi Darshan" programme in India is telecasted from — **New Delhi**
- ☆ The weekly agriculture column of "The Hindu" appears on — **Thursday**
- ☆ The DRDA was established in — **1980**
- ☆ The theory of human motivation is given by — **A. H. Maslow**
- ☆ PERT and CPM is — **Management technique**
- ☆ Brain storming method was developed by — **Osgood**
- ☆ The learning curve in training follows a — **S-shape**
- ☆ The apex body to offer training for extension functionaries — **MANAGE**
- ☆ Element or Function of Management According to the Luther Gullick — **POSDCORB**
- ☆ Adult learning is called — **Andragogy**
- ☆ Child learning is called — **Pedagogy**
- ☆ Adult leaning is — **Problem centered**
- ☆ World literacy day — **8 September**
- ☆ National women literacy day — **2 October**

☆ Father of Rural Sociology — **Auguste Compte**

☆ Family is a — **Primary group**

☆ Basic unit of development under IRDP is — **Family**

☆ Basic unit of civilization — **Family**

☆ Basic unit of rural society is — **Village**

☆ Who firstly gave the idea of folkways — **W.G. Sumner**

☆ Kinesics is the study of — **Body language**

☆ NFSM aims to increase food grain product of — **Wheat, Rice and Pulse**

☆ The effective rate of delivering a radio talk for extension purpose — **120 – 140 words/min.**

☆ Namma Dhawani is a — **Community radio**

☆ Thematic Appreception Test (TAT) is used to measure — **Attitude**

☆ Gate keeping theory was given by — **Kurt Lewin**

☆ Research is a tool for measurement of — **Intelligence**

☆ Which country launched its first extension program named Harvested Programme — **Israel**

☆ Farm magazine Prasardoot is published from — **IARI, New Delhi**

☆ Age of rural youth in TRYSEM is — **18 – 35 years**

☆ First department of agriculture was established in — **1881**

☆ Film projector commonly used in extension — **16 mm**

☆ A3P launched in 2010-11 under — **NFSM**

☆ When two or more brief talks presenting phases of the same general topic is called — **Symposium**

☆ Demonstration conducted by farmer under the direct supervision of extension worker is a — **Result demonstration**

☆ Demonstration conducted to show the technique of doing things is — **Method demonstration**

☆ Demonstration conducted by research worker in farmer field is — **Front line demonstration**

☆ Principle of extension education is — **Learning by doing**

☆ Goal of extension education is — **To promote income of farmers**

☆ Fundamental objective of extension education is — **Development of people**

☆ Purpose of extension evaluation is to identify the — **Weak points, strong points, gaps and errors**

☆ Success of rural development project depends upon — **Participation of beneficiaries**

☆ Lecture, Group discussion, Seminor, Symposium, Syndicate, Pannel discussion, Case study, Role play, Brain storming, Buzz session are the method of — **Training**

☆ For study of farming system, the best PRA exercise is — **Resource mapping**

☆ T and V system is a good example of — **Training approach**

☆ News paper, Demonstration, Symposium are example of — **Group communication**

☆ Television, Film, Tape recorder are example of — **Mass communication**

☆ A working model is known as — **Mock-up**

☆ Best method of leader selection — **Sociometry**

☆ Total population in India in villages — **3/4**

☆ Formula of Intelligence Quotient (I.Q.) — **(Mental age/ Chronological age)×100**

☆ State Agriculture Universities in India were in the pattern of — **Land Grant Colleges (USA)**

☆ Farmers which are first to adopt a new idea — **Innovators**

☆ Farmers which are last to adopt a new idea — **Leggards**

☆ **Adopter categories**

  ☆ Innovator – (Venturesome) – 2.5 per cent

  ☆ Early adopter – (Respectable) – 13.5 per cent

  ☆ Early majority – (Deliberate) – 34 per cent

  ☆ Late majority – (Sceptical) – 34 per cent

  ☆ Laggards – (Traditional) – 16 per cent

☆ **Programme planning process according to Person (1966)**

  i) Collect Facts

  ii) Analysis of the situation

  iii) Identify the problem

  iv) Decide the objective

  v) Develop plan of work (5W and 1H)

  vi) Execute the plan

vii) Evaluation or review of the process

viii) Reconsideration.

☆ **Programme planning process according to J. P. Legans**

i) Situations and Problems

ii) Objectives and Solutions

iii) Teaching plan of work

iv) Evaluation

v) Reconsideration

21
Agricultural Economics

☆ Father of economics   – **John Neville Keynes**

☆ Father of agricultural economics   – **Adam Smith**

☆ Father of co-operative movement in India was   – **F. Nicholson**

☆ Population theory was proposed by   – **Malthus**

☆ FCI was established in   – **1965**

☆ NAFED was established in   – **1958**

☆ Warehousing corporation act come   – **18th March 1962**

☆ WTO established in   – **1995**

☆ First director general of WTO   – **Renato Ruggiero**

☆ Agricultural produce act was passed in   – **1937**

☆ AGMARK act was passed in   – **1937**

☆ AGMARK lab situated at   – **Nagpur**

☆ AGMARK is an indicator of   – **Purity**

☆ 1st bank in India was established in   – **1806**

☆ National Institute of Agriculture Marketing (NIAM) is situated at   – **Jaipur**

☆ Headquarter of World bank   – **Washington D.C**

☆ RBI was established in   – **1st April 1935**

☆ RBI was nationalized in   **– 1ˢᵗ January 1949**

☆ First governor of RBI   **– Osborne Smith**

☆ 14 commercial banks was nationalized in   **– 19ᵗʰ July 1969**

☆ Agriculture year   **– 1ˢᵗ June to 31ˢᵗ May**

☆ Financial year   **– 1ˢᵗ April to 31ˢᵗ March**

☆ Full form of NABARD   **– National Bank for Agricultural and Rural Development**

☆ NABARD was established in   **– 12 July 1982**

☆ NABARD was set up on the recommendation of   **– B. Sivaraman Committee**

☆ Export Import bank of India was established in   **– 1 January 1982**

☆ First mortgage bank (Land development bank) was established in   **– Punjab (1920)**

☆ First of all 5 RRBs were established in   **– Moradabad, Gorakhpur, Bhiwani, Jaipur, Malda**

☆ Theory of absolute advantage was given by   **– Adam Smith**

☆ Theory of profit was given by   **– Walker**

☆ Theory of inflation was given by   **– Lerner**

☆ Wage fund theory was given by   **– Mill**

☆ Theory of undeveloped countries was given by   **– Keynes**

☆ Modern theory of wages and employment was given by   **– Keynes**

☆ Theory of capital was given by   **– Karl Marks**

☆ Theory of rent was given by   **– Ricardo**

☆ Inflation means   **– Persistence rise in general price level**

☆ Inflation is measured by   **– WPI**

☆ Price theory is branch of   **– Micro-economics**

☆ Aggregate supply of labour in case of classical is   **– Vertical line**

☆ Economics concerned with individual unit *i.e.* single industry   **– Micro-economics**

☆ Economics deal with whole economics setup *i.e.* total production, total expenditure   **– Macro-economics**

☆ Process of estimating costs, returns and net profit of a farm or a particular enterprise   **– Farm budgeting**

☆ Most effective way to overcome the defects of
agriculture marketing is — **Regulating marketing**

☆ Those product, in which two products produced
together are called — **Joint product**

☆ Expense to income ratio is called income to expensection — **Efficiency ratio**

☆ Basis of farm budgeting is — **Cost benefit analysis**

☆ Net worth of farm business is calculated from — **Balance sheet**

☆ Statement which provides repayment schedule of
farmer is — **Cash flow statement**

☆ Point at which total cost and total revenue is equal to — **Break even point**

☆ When farm is classified on the basis of utilization of
land and resources is called — **Types of farming**

☆ Farming having >50 per cent income by
single enterprise — **Specialized farming**

☆ Crop production + livestock raising is called — **Mixed farming**

☆ In mixed farming, the contribution of livestock to
gross farm income is at least — **10 per cent**

☆ Farming which has <50 per cent income by
single enterprise — **Diversified farming**

☆ Natural grazing pattern is known as — **Ranching**

☆ When farm is classified on the basis of organizational
setup is called as — **System of farming**

☆ Joint agriculture operation by farmer on
voluntary basis — **Cooperative farming**

☆ Farming in which investment of land and capital is
done by big business person or capitalist — **Capitalistic farming**

☆ When farmers follows agriculture practices in their
own way and managers and organizer of their
farm business — **Peasant farming**

☆ Government carries out farming is — **State farming**

☆ Process of deciding in the present what to do in the
future about the best combination of crops and
live stock to be raised — **Farm planning**

☆ Dumping activity is seen under — **Monopoly**

☆ When there is a single seller of a product — **Monopoly market**

☆ Market consisting of single buyer of a product — **Monosomy market**

- ☆ When few sellers of commodity — **Oligopoly market**
- ☆ When few buyers of a commodity — **Oligopsony market**
- ☆ A sustained rise of less than 3 percent of price per annum is called — **Creeping inflation**
- ☆ Contribution of central government in capital share of Regional Rural Bank — **50 per cent**
- ☆ Portion which is usually brought to market at a particular time for sale — **Market surplus**
- ☆ Relationship between marketable to marketed surplus for perishable products — **Equal**
- ☆ Minimum Support Price (MSP) fixed by — **Commission of Agriculture Cost and Price (CACP)**
- ☆ MSP and procurement price is announced by — **GoI**
- ☆ CACP recommended MSP for How many crops — **25**
- ☆ Repayment capacity of loan of a farmer is judged on the basis of — **Total cropped land**
- ☆ Short term loan given for the period of — **1 to 1.6 years**
- ☆ Long term loan given for the period of — **5 to 30 years**
- ☆ 3R of credit are — **Return, Repayment capacity, Risk bearing ability**
- ☆ Increase in money supply and fall in production causes — **Inflation**
- ☆ Money supply in Indian national economy is regulated by — **RBI**
- ☆ Income tax is example of — **Direct tax**
- ☆ Goods, purchases, services product tax is example of — **Indirect tax**
- ☆ Multistage sales tax with credit for taxes paid on business purchase — **VAT**
- ☆ VAT (Value Added Tax) is a — **Indirect tax**
- ☆ Slope of iso-cost line is — **Downward**
- ☆ Land holding of marginal farmers — **<1 ha area**
- ☆ Land holding of small farmer — **1-2 ha area**
- ☆ Land holding of large farmer — **>2 ha area**
- ☆ Zamindari system was introduced by — **Lord Cornwallis**
- ☆ Which basis, CACP fixed minimum prices of crops — **Cost of production**
- ☆ Production function is also known as — **Input output relation**

- ✩ Production is a function of — **Factor**
- ✩ Cost calculation per hectare is also known as — **Cost of production**
- ✩ Fixed cost is also known as — **Overhead cost**
- ✩ Most important unit of farm management — **Production unit**
- ✩ Farm machinery and equipment are example of — **Working assets**
- ✩ Prime cost is also called — **Variable cost**
- ✩ Opportunity cost also called — **Alternative cost**
- ✩ Marginal word in economics means — **Additional**
- ✩ When demand and price are equal is called — **Equilibrium price**
- ✩ Basic fundamental law of agriculture is — **Law of diminishing return**
- ✩ Diminishing return is also called — **Law of variable**
- ✩ Law of variable proportion is generally referred to as — **Law of diminishing return**
- ✩ Theory of rent is based on — **Law of diminishing marginal returns**
- ✩ When two products are substituting each other, the optimum level of these enterprises can be worked out through — **Production possibility curve**
- ✩ Basis of Cobb Douglas production function is — **Constant elasticity of substitution**
- ✩ Under perfect competition market, maximum profit is obtained when — **Marginal return = Marginal cost**
- ✩ Net operation income is equal to — **Gross income – (Operating expenses + Depreciation on working assets)**
- ✩ Net farm income is equal to — **Net operating income – (Fixed expenses + Depreciation on fixed assets)**
- ✩ Current ratio is equal to — **Total current assets/ Total current liabilities**
- ✩ Current liability ratio is equal to — **Current liability/Owner equity**
- ✩ Net cash income is equal to — **Total cost receipt – Total operating cost**
- ✩ Operating cost ratio is equal to — **Total operating cost/Total profit**
- ✩ Fixed ratio is equal to — **Fixed expenses/Gross income**

- ✩ Gross ratio is equal to — **Total expenses/Gross income**
- ✩ Marginal cost is equal to — **Change in total cost/ Change in output**
- ✩ Net capital ratio is equal to — **Total assets/Total liabilities**
- ✩ Variable cost is equal to — **Prime cost/Input cost**
- ✩ Average cost is equal to — **Total cost/Out put**
- ✩ Total cost is equal to — **Fixed cost + Variable cost**
- ✩ Profit is equal to — **Gross income – Total cost**
- ✩ Net income equal to — **Gross return - Cost**
- ✩ All actual expenses in cash and kind incurred in production by owner operator — **Cost A_1**
- ✩ Cost A_1+ rent paid for based in land — **Cost A_2**
- ✩ Cost A_2 + interest on value of owned capital assets (including land) — **Cost B_1**
- ✩ Cost B_1+ rental value of owned land and rent paid by leased in land — **Cost B_2**
- ✩ Cost B+ imputed value of family human labour — **Cost C**
- ✩ Total cost of production which includes all cost items, actual as well as imputed — **Cost C**
- ✩ Cost C is also called — **Gross cost or Total cost of cultivation**
- ✩ Gross return — **Cost A is – Farm business income**
- ✩ Gross return — **Cost B is – Family labour income**
- ✩ Gross return — **Cost C is – Net income**
- ✩ Benefit cost ratio is — **Gross income/Cost C**
- ✩ Land rent is example of — **Fixed cost**
- ✩ Optimum profit will be obtained at a point where — **MC=MP**
- ✩ When MP=0, then Ep=0 is called — **Completely inelastic demand**
- ✩ When MP>AP then Ep>1 is called — **Elastic demand**
- ✩ When MP=AP, then Ep=1 is called — **Unit inelastic demand**
- ✩ Ep means — **Elasticity of production**
- ✩ MP — **Marginal production**
- ✩ TP — **Total production**

Important Relations

☆ When MPP>APP then APP is — Increasing

☆ When MP is 0 then TP is — **Maximum**

☆ When TP is maximum than MP is – **Zero**

☆ When TP is zero than AP is – **Zero**

☆ When MP is increasing than AP is – **also increasing**

☆ When AP is maximum than – **MP=AP**

☆ When AP is increasing than MP is – **greater than AP**

☆ When AP is decreasing than - MP is – **less than AP**

☆ When MP is rising than – **TP increasing at a increasing rate**

☆ When AP is increasing than – **TP is increasing at a decreasing rate after MP at its maximum**

22
Agricultural Statistics

☆ Father of statistics   **– R. A. Fisher**

☆ Kurtosis term was introduced by   **– Karl Pearson (1906)**

☆ Student t test was given by   **– W. S. Gosset (1908)**

☆ Student t test was perfected by   **– R. A. Fisher (1926)**

☆ ANOVA technique developed by   **– R. A. Fisher**

☆ x^2 test was given by   **– Karl Pearson (1990)**

☆ SD term was first used by   **– Karl Pearson (1894)**

☆ Binomial distribution was given by   **– James Bernoulli (1700)**

☆ First step of summarizing the data is   **– Classification**

☆ Represent the whole data by only single value is called   **– Measure of central tendency**

☆ Mean, Median, Mode are   **– Measure of central tendency**

☆ A figure obtained by dividing the sum of all variable by their total number of variable   **– Arithmetic mean (A.M.)**

☆ Find the average height of plants and find average speed when time for each speed is fixed measured by   **– Arithmetic mean (A.M.)**

☆ Best measures of central of tendency   **– A.M.**

☆ Sum of deviation of items from the A.M. is — **0**

☆ Which mean is affected by change in origin and scale — **A.M.**

☆ Sum of square of deviation of the variate from their mean is — **Least**

☆ Middle most value of the series is — **Median**

☆ Positional average is — **Median**

☆ Which one is represent median — **50th Percentile**

☆ Most frequently occurred item — **Mode**

☆ Find average size of shoes sold in the market we should use — **Mode**

☆ 3 Median – 2 Mean is equal to — **Mode**

☆ Ratio of number of observations to the sum of the reciprocal of the value of the different observation — **Harmonic mean (H. M.)**

☆ Mean applied when deals with rate, price and speed of a vehicle — **H. M.**

☆ HM is reciprocal of — **A.M.**

☆ Measure of central tendency to used to study average rate of change in population is — **Geometric mean (G. M.)**

☆ Mean applied when deals with relative changes *e.g.* Bacterial growth, cell division, population — **Geometric mean (G. M.)**

☆ Average of the sum of squares of the deviation about mean — **Variance**

☆ Square of SD is — **Variance**

☆ Degree of scatter ness or variation of the variable about a central tendency — **Dispersion**

☆ MD, SD and Variance are — **Measures of dispersion**

☆ Best measure of dispersion is — **Standard deviation (S.D.)**

☆ If all variate value are negative the SD will be — **Positive**

☆ SD is always calculated by — **A.M.**

☆ SD is ranges from — **0 to α**

☆ Unit less figure based on two value — **Range**

☆ Difference between highest and lowest value of the series — **Range**

☆ Quality control in industries is the most important measure of dispersion — **Range**

☆ When mean is 97, variance is 81 then CV is — **9.27**

- ☆ Coefficient of variation (C.V.) is calculated by — **SD/Mean×100**
- ☆ Variation used to compare the variability between two series — **C. V.**
- ☆ ½ of the inter quartile range is — **Quartile deviation (Q. D.)**
- ☆ Relation between AM, GM and HM is — **AM > GM > HM**
- ☆ Measure of the direction and degree of asymmetry — **Skewness**
- ☆ When Mean > Median > Mode, skewness will be — **Positive**
- ☆ When Mean, Median, Mode is equal than skewness will be — **Zero**
- ☆ Left skewed is also known as — **Negative skew**
- ☆ Right skewed is also known as — **Positive skew**
- ☆ Graphical display of tabulated frequencies, shown as bars, is known as — **Histogram**
- ☆ Idea about the flatness/peakedness of the curve — **Kurtosis**
- ☆ β_2 is measure of — **Kurtosis**
- ☆ Curve have $\beta_2 > 3$ or $Y_2 > 0$ is — **Leptokurtic curve**
- ☆ Curve have $\beta_2 = 3$ or $Y_2 = 0$ is — **Mesokurtic curve**
- ☆ Formula of Karl pearson's coefficient of skewness — **(Mean-Mode)/σ**
- ☆ Coefficient of skewness for normal distribution is — **0**
- ☆ Study the association or degree and deviation between two or more variable — **Correlation**
- ☆ Correlation coefficient lies between — **-1 to +1**
- ☆ Correlation is independent of change of — **Origin and Scale**
- ☆ Which is used to measure the average relationship between two or more variables — **Regression**
- ☆ Regression coefficient is independent of change of — **Origin**
- ☆ Regression coefficient lies between — **- α to + α**
- ☆ Regression line is sometimes called — **Line of best fit**
- ☆ Mean, Median, Mode are equal in — **Normal distribution**
- ☆ Normal distribution has a skewness — **Zero**
- ☆ Laplace used the normal distribution in — **Analysis of error of experiments**
- ☆ Normal distribution is also known as — **Law of error or Gaussian law**
- ☆ Distribution in which Mean > Variance — **Binomial distribution**
- ☆ Mean of the binomial distribution is — **np**

- ☆ SD of binomial distribution — $\sqrt{npq}$
- ☆ Standard error calculated by — $\sigma/\sqrt{n}$
- ☆ Distribution in which Mean = Variance — **Position distribution**
- ☆ Degree of freedom of Normal distribution — **n-3**
- ☆ Term used to denote chance of happening or not happening of an event — **Probability**
- ☆ Probability is science of — **Decision**
- ☆ Probability of impossible event is — **Zero**
- ☆ Formula of Probability — **No. of favourable cases/ Total no. of equally likely cases**
- ☆ Probability ranges from — **0 to 1**
- ☆ If events, A and B are independent then the joint probability is — $P(A\cap B)=P(A \text{ and } B)=P(A).P(B)$
- ☆ If two events are mutually exclusive then the probability of either occurring is — $P(A \text{ or } B)=P(A\cap B)=P(A)+P(B)$
- ☆ Conditional probability $P(A/B)$ is equal to — $P(A\cap B)/P(B)$
- ☆ When 4 dice are thrown, what is the probability that the same number appears on each of them — **1/216**
- ☆ Outcome or set of out come associated with a certain condition is — **Event**
- ☆ Different possible result of an experiment is — **Outcome**
- ☆ Events which occur only once or will exclude the occurrence of other is called — **Mutually exclusive event**
- ☆ Occurrence of one events does not affect the occurrence of other is called — **Independent event**
- ☆ Which test used for comparing two means when sample size is small (up to 30) — **T test**
- ☆ T test some time called as — **Parametric test**
- ☆ Student t test is used when — **Small sample size and SD is unknown**
- ☆ To test the proportions and variance we use — **F test**
- ☆ x^2 is a index of — **Dispersion**
- ☆ To test the goodness of fit or homogeneity we use — **x^2 test (Chi-square test)**
- ☆ x^2 lies between — **0 to α**

✩ Test the agreement between observed frequencies
and expected frequencies we use   – χ^2 **test**

✩ For testing the independence of two attributes, the test used is   – χ^2 **test**

✩ Minimum sample size for using x^2 test should be   **– 50**

✩ Distribution of x^2 depends on the   **– Degree of freedom**

✩ When the calculated F is greater than table F value at
5 per cent only, the differences in treatment is considered   **– Significant**

✩ With increasing number of error degree of freedom,
table F value follow   **– Gradually decreased trend**

✩ Basic principles of field experimentation   **– Replication, Randomization and Local control**

✩ Repeated application of treatments   **– Replication**

✩ Allocation of treatments to the different experimental
units by a random process   **– Randomization**

✩ Which principle of experimentation eliminates human
biases   **– Randomization**

✩ Local control helps in reducing   **– Experimental error**

✩ Full form of ANOVA   **– Analysis of Variance**

✩ Correlation of continuity in 2×2 contingency table
should be used when expected frequency of any cell is   **– Less than 5**

✩ Hypothesis which is under test for possible
rejection is   **– Null hypothesis (Ho)**

✩ Alternate hypothesis is also known as   **– Research or maintained hypothesis**

✩ Error in which hypothesis is true but our
test rejects it   **– Type I error (Rejection error)**

✩ Out of the two types of error in testing, the
more severe error is   **– Type II error (Acceptance error)**

✩ Rejection of null hypothesis when it is false
is known as   **– Type II error**

✩ Type II error is   **– More serious then Type I error**

✩ Simplest experimental design   **– Complete Randomized Design (CRD)**

✩ Experimental design which provides maximum
degree of freedom for error   **– CRD**

✩ Which design is applied when experimental material
are limited and homogenous   **– CRD**

☆ Error degree of freedom in CRD is formulated as — **N-t**

☆ CRD is frequently used in — **Laboratory experiments**

☆ Most commonly used design — **Randomized Block Design (RBD)**

☆ When fertility gradient in one direction, the statistical design to be used — **RBD**

☆ Maximum number of treatments adopted in RBD — **20**

☆ Degree of freedom for error in RBD with 10 treatments and 4 replication will be — **27**

☆ One way elimination of heterogeneity design/Two way classification of ANOVA is also called as — **RBD**

☆ Error degree of freedom of RBD is formulated as — **(t-1)(r-1)**

☆ Design in which fertility gradient is in two way direction — **Latin Square Design (LSD)**

☆ Two way elimination of heterogeneity design/ Three way classification of ANOVA is also called as — **LSD**

☆ In LSD, the number of row or column or treatment is equal to — **Number of replication (r=c=t)**

☆ Optimum number of treatments studied in LSD — **5 to 12**

☆ Error degree of freedom of LSD is formulated as — **(t-2)(t-1)**

☆ Which design provides main effects and interaction — **Factorial RBD**

☆ Most appropriate design, when all factors are not of equally important in experimentation — **Split Plot Design (SPD)**

☆ Factor requires larger units to be applied and may produce larger differences — **Main plot**

☆ Error degree of freedom of SPD is formulated as — **D(r-1)(d-1)**

☆ If sub treatment are laid out in strips then the design is called — **Strip plot design**

☆ How many number of error variance are applied in strip plot design — **3**

☆ Critical region is also called as — **Region of rejection**

☆ Generalize mean is also known as — **Holder mean**

23
Veterinary and Animal Husbandry

☆ The world`s first *in-vitro* fertilized buffalo calf — **Pratham**

☆ Female over 1 year age and is yet to give birth to a calf is — **Heifer**

☆ Milk production is maximum in India — **U.P.**

☆ AICRP on poultry — **Izatnagar (1970)**

☆ NDRI situated at — **Karnal**

☆ Buffalo, Goat and Sheep belong to the family of — **Bovidae**

☆ Cow`s milk is light yellow or creamy in colour due to — **Carotene**

☆ Cow`s milk is white in colour due to presence of — **Casein**

☆ Weaning of cow calf is done at — **After one month of age**

☆ Casein content of cow milk —**3.0 per cent**

☆ Casein content of buffalo milk — **4.3 per cent**

☆ Lactose content of cow milk — **4.5 per cent**

☆ Lactose content of buffalo milk — **4.8 per cent**

☆ Highest and sweetest milk producing cow breed — **Sahiwal**

☆ Highest milk producing hybrid cow breed — **Holstein Frieswal (14 lit/day)**

☆ Highest body weight in Indian cow — **Kankrej**

☆ Highest fat per cent in exotic breeds of cow — **Jersey**

☆ Wallowing is a common behaviour of — **Buffalo**

☆ Highest milk fat per cent found in breed of buffalo is — **Bhadawri (8-13 per cent)**

☆ Highest milk producing buffalo breed — **Murrah**

☆ Respiration rate of buffalo — **15-20 per minute**

☆ Heat period of cow and buffalo is — **18-36 Hrs**

☆ Oxytocin hormone responsible for — **Milk let down**

☆ Milk is deficient in — **Iron**

☆ Boiling point of milk — **100.17 °C**

☆ pH of pure milk — **6.5-6.8**

☆ specific gravity of cow milk is — **1.03**

☆ Average acidity of freshly drawn milk is — **0.14-0.18 per cent**

☆ How much moisture percentage in cow ghee — **0.2 per cent**

☆ NaCl quantity in good milk — **0.12 per cent**

☆ Ring worm disease is caused by — **Fungus**

☆ Milk is very good source of — **Calcium and Phosphorus**

☆ First clone of adult animal — **Dolly**

☆ Goat is called — **Poor man`s cow**

☆ Adult male of goat is called — **Buck**

☆ Maximum milk fat found in breed of goat — **Black bengal**

☆ Mohair produced by which animal — **Angora breed of Goat**

☆ Dolly clone of sheep is created by — **Wilmont**

☆ Docking in sheep is done at the age of — **7-14 Days**

☆ New born of sheep is known as — **Lamb**

☆ Remove of tail in sheep is called — **Docking**

☆ Castrated male sheep is called — **Wedder**

☆ Castrated male pig is called — **Hog**

☆ Uncastrated adult male pig is called — **Boar**

☆ Castrated male horse is called — **Geld**

☆ Salted smoked meat of pig is known as — **Bacon**

☆ Morocco leather a fine leather is prepared from — **Goat skin**

☆ Egg shell is made of — **$CaCO_3$**

☆ Quality of egg can be judge by — **Candling**

☆ NRC for Horse is situated at — **Hissar**

☆ Number of recognized breeds of goat in India is — **20**

☆ Processing of exposing milk about 61-63 °C for 30 min or 71-76 °C for 15 second is called — **Pasteurization**

☆ Livestock census is done after every — **4 Years**

☆ Mating of unrelated pure breed animals within the same breed is called — **Out crossing**

☆ Mating of animals of different breeds known as — **Cross breeding**

☆ Largest part of ruminant stomach — **Rumen**

☆ Third compartment of ruminant stomach is known as — **Omasum**

☆ Livestock insurance scheme was started during — **2005-06**

☆ Animal insurance scheme was started from — **1974**

☆ National fisheries development board situated in — **Hyderabad**

☆ National fisheries development board established in — **September 2006**

☆ Castrated male of pigs is called — **Hog**

☆ Testosterone is also known as — **Male hormone**

☆ Digestive juices are secreted from — **Abomasum part of ruminant stomach**

☆ Rinder pest is also known as — **Cattle plague**

☆ Anthrax is also known as — **Spleenic fever**

☆ Fat percentage of salted butter is — **80 per cent**

☆ Minimum dry period of cow should be — **40 days**

☆ Plastic cream should contain — **45-85 per cent milk fat**

☆ Temperature of artificial vagina — **42°C**

☆ In cow/buffalo the peak milk yield is obtained at — **4-8 weeks after parturition**

☆ According to PFA rules the toned milk should contains fat percentage not less than — **3 per cent**

☆ Number of well defined indigenous cattle breeds in India is — **28**

☆ Best breed of buffalo in the world is — **Murrah**

☆ Saanen is called — **Milk queen of goat world**

☆ Milk fever disease is due to deficiency of — **Ca**

☆ Urea feeding in ruminants is due to the supply of — **Protein**

☆ Artificial insemination programme was first started in India at the place of — **Dairy farm at Mysore (1939)**

☆ Milk fat contain Cholesterol, while vegetable fats contain — **Phytosterol**

☆ Average life of cow and buffalo are — **18 years**

☆ Milk is also called — **Complete food**

☆ First milk produced by female is called — **Colostrum**

☆ Process of eliminating non productive animal is called — **Culling**

☆ Perosis in cattle occur due to the deficiency of — **Manganese**

☆ Which animal resistant to tuberculosis is — **Goat**

☆ Queen of leguminous green fodder is — **Berseem**

☆ Best non leguminous winter green fodder for milch animals is — **Oats**

☆ First vaccination against Haemorrhagic septicaemia in cattle should be done at the age of — **4-6 month**

☆ Most dwarf breed of cattle is — **Vechur**

☆ Best method of age determination in buffalo is by — **Dentition**

☆ Operation flood-II was launched in — **1978**

☆ Major portion of milk fat is in form of — **Triglyceride**

☆ For making good silage the dry matter percentage in green fodder should be — **20-25 per cent**

☆ Hormone responsible for holding up of milk is — **Adrenaline**

☆ Purity of semen detected by — **BBC test**

☆ Popular egg producing breed of poultry is — **Leghorn**

☆ Marek`s disease is found in — **Poultry**

☆ Domestic fowl is called — ***Gallus domesticus***

☆ Cochin breed of poultry is also known as — **Shanghai fowl**

☆ Hatching time period in fowl is — **21 days**

☆ Which breed of poultry is originated from Dhar district of M.P. — **Karaknath**

☆ New castle disease is also known as — **Ranikhet disease**

☆ Chittagong breed of poultry is also known as — **Malay**

☆ 8-22 weeks of age chicken is called — **Grower**

☆ More than 22 weeks old chicken is called — **Layer**

☆ Male fowl below one year of age is called — **Cockaral**

☆ Mature male fowl is known as — **Cock**

☆ Cow breeds — **Karan fries (Holstein friesian ×Tharparkar), Karan swiss (Brown swiss × Sahiwal or Red sindhi), Frieswal (Holstein × Sahiwal)**

☆ Cow breeds for milch purpose — **Sindhi, Sahiwal, Gir, Deoni**

☆ Cow breeds for dual purpose — **Tharparkar, Kantkrej, Haryana, Nimari, Gaolao, Krishna Valley, Ongole**

☆ Cow breeds for draught purpose — **Hallikar, Amrit mahal, Dangi, Mewati, Hissar, Kangyam, Ganga titri**

☆ Exotic breed of cattle — **Holstein Friesian, Brown swiss, Jersey, Ayreshire**

☆ Buffalo breeds — **Murrah, Jafrabadi, Surti, Nili ravi, Bhadawari, Parlakiddi, Mehsana, Godavari, Toda**

☆ Goat breeds — **Angora, Jamunapari, Barbari, Surathi, Alpine, Sirohi, Chegu, Marwari, Mehsana**

☆ Sheep breeds — **Nellore, Rampur bushair, Deccan, Marino,**

☆ Pig breeds — **Large white Yorkshire**

☆ Triple purpose breed of sheep is — **Lohi**

☆ Heavy breed of cow is — **Kankrej**

☆ Dual purpose breed of goat is — **Jamunapari**

☆ Sussex breed of poultry developed in — **England**

☆ Breed of poultry — **Karaknath, Cochin, Brahma**

24
Agricultural Engineering

- ☆ Soil strength is determined by — **Penetrometer**
- ☆ Voltage in a spark plug at the time of spark is — **20000 V**
- ☆ Break testing is done at — **25 kmph**
- ☆ Puddling is done to — **Reduce percolation of water**
- ☆ Piezometer is used to measure — **Static pressure of flowing fluid**
- ☆ Air cooling system is generally found in — **Single cylinder engine**
- ☆ Cold spark plug is used on — **Heavy engine**
- ☆ Reaper is a used for — **Cutting crop**
- ☆ V shaped sweeps are best suited for — **Stubble mulch tillage**
- ☆ Large tilt angle is best for — **Sticky and non scouring soil**
- ☆ Chock in carburettor is provide to — **Control air supply**
- ☆ Chock controls the amount of — **Air**
- ☆ Plough used for maximum moisture conservation is — **Chisel plough**
- ☆ Most mechanized crop in India — **Wheat**
- ☆ Specific gravity of fully charged battery is — **1.28**
- ☆ Sub soiler plough is best suited for — **Breaking hard fan**
- ☆ Sub soiler are operated at maximum depth of — **45-75 cm**
- ☆ Number of piston rings on piston varies between — **3-7**

☆ Reciprocal type mower is fitted with — **Rotary blade**

☆ Disk plough are used when the soil is — **Tough**

☆ Fly wheel type chaff cutter used — **Worm gear**

☆ Horizontal plate seed metering device is used in — **Planter**

☆ Scalper is used for — **Removing of stone**

☆ Specific fuel consumption of diesel engine is — **Less than in petrol engine**

☆ Main combustible constituent of bio gas is — **Methane**

☆ Average force that a bullock can expert — **1/10th of their body weight**

☆ Cyclone separator is used for separating the — **Fine particles**

☆ Vertical suction of a plough influences — **Depth of cut**

☆ In an offset harrow, the number of gang is — **2**

☆ Belt tension can be adjusted by — **Idler pulley**

☆ Smallest size of economical gobar gas plant is — **70 cu. Ft**

☆ Indigenous plough is — **Primary tillage implement**

☆ Length of cutter bar for the tractor operated mower or vertical conveyor reaper is — **1.5-3 m**

☆ Radius of curvature of a disc with 60 cm dia, 10 cm concavity will be — **60 cm**

☆ Which stage represents optimum conditions for tillage — **Crumby**

☆ Combine harvestor with 4 m cutter bar and 4 km/hr speed of operation will harvest — **1.6 ha/hr**

☆ Most suitable furrow opener for seeding in a trashy and hard seed bed is — **Disc type**

☆ Working life of a tractor drawn cultivator is usually — **5000 hrs**

☆ Tractor seat suspension should have its natural frequency in the range of — **2-4 cpc**

☆ Negative slip of tractor is obtained in field operation of — **Rotatilling**

☆ Normal consumption of fuel in litres/day by a 35 hp tractor is — **3**

☆ Specific fuel consumption of tractor diesel engine is fron — **0.18-0.25 kg/bph-hr**

☆ Compression pressure in petrol engine cylinder is — **6-10 kg/cm^3**

☆ Injection pressure of the fuel nozzle in diesel engine is — **100-200 kg/cm^3**

☆ Inflation pressure in rear wheel of tractor is — **0.8-1.2 kg/cm^3**

☆ Separation of liquids from solids by the application of pressure is known as — **Expression**

☆ Power tiller is most suitable for the cultivation of — **Paddy**

☆ Popular power tiller used in Indian agriculture are in the power range of — **9-11 KW**

☆ Water requirement of any crop is measured in — **cm of water**

☆ If size of a seed drill is doubled and speed is halved, then coverage will be — **Unchanged**

☆ Most commonly used pumps in tractor hydraulic system are — **Gear type**

☆ Gears commonly used on fuel injection pumps are — **Rack and pinion**

☆ Gear generally used in transmission box are — **Helical and Spur gears**

☆ Lowest temperature at which the fuel ceases to flow is known as — **Pour point**

☆ Desi plough, MB plough, Ridge plough, Disk plough are — **Primary tillage implements**

☆ Cultivator, Harrows, Hoe, Roller, Spike tooth are — **Secondary tillage implements**

☆ Disc plough is used for — **Deep ploughing in grass field**

☆ Harrow is used for — **Preparation of seedbed, destroy weeds**

☆ Rotary plough is used for — **Cut and pulverizes the light soil**

☆ Star weeder is used for — **Weeding in dry lands and groundnut fields**

☆ Tillage implement used to break sub soil is — **Chisel plough**

☆ Tillage implement used to Earthingup, form ridge and Furrows is — **Ridge plough**

25
Fisheries

- ☆ ICAR-Central Institute of Marine Fisheries Research situated at — **Kochi (Kerala)**

- ☆ ICAR-Central Inland Fisheries Research Institute (CIFI) situated at — **Barrackpore (W.B.)**

- ☆ ICAR-Central Institute of Fisheries Technology situated at — **Kerala**

- ☆ ICAR-Central Institute of Coastal Engineering for Fishery situated at — **Karnataka, Bangalore**

- ☆ Main freshwater fish are — **Carp and Catfish**

- ☆ Main brackish water fish are — **Hilsa and Mullet**

- ☆ Blue revolution is related to — **Fish production**

- ☆ Study of fishes is called — **Ichthyology**

- ☆ Growing plants in water is called — **Aquaculture**

- ☆ Study of fish husbandry is called — **Pisiculture**

- ☆ ICAR- Central Institute of Brackishwater Aquaculture, Chennai, Tamilnadu.

- ☆ ICAR- Central Institute of Fisheries Education, Mumbai, Maharashtra.

- ☆ ICAR- Central Institute of Fisheries Technology, Kochi, Kerala.

- ☆ ICAR- Central Institute of Freshwater Aquaculture, Bhubaneshwar, Odisha.

- ☆ ICAR- National Bureau of Fish Genetic Resources, Lucknow, Uttar Pradesh.

- ☆ ICAR- National Research Centre on Coldwater Fishries, Bhimtal, Uttarakhand.

Exam Oriented Information

- ☆ Total Geographical Area — 328 Mha (3287263 sq.km)
- ☆ Highest non-irrigated area in India — Mizoram
- ☆ Net Cropped Area — 143 Mha
- ☆ Gross irrigated area — 76 Mha
- ☆ Net irrigated area (2016-17) — 64.7 Mha
- ☆ Micro irrigation Area in India — 4 Mha
- ☆ Drip irrigation Area in India (2016-17) — 3.37 Mha
- ☆ Sprinkler irrigation Area in India (2016-17) — 4.36 Mha
- ☆ Availability of land in India (2010-11) — 0.12 ha
- ☆ Highest area under drip irrigation in India — Maharashtra
- ☆ Highest area under sprinkler irrigation in India — Haryana
- ☆ Highest net irrigation potential under Canal and Tubewell in India — U.P.
- ☆ Gross cropped area — 193 mha
- ☆ Area sown more than once — 50 mha
- ☆ Average annual rainfall — 1190 mm
- ☆ Average Cropping intensity of India — 135 per cent

☆ National Center of Organic Farming is situated at — **Ghaziabad (U.P.)**

☆ National Seed Research and Training Center is situated at — **Varansi (U.P.)**

☆ Rajendra Prasad Agricultural University was upgraded to — **Dr. Rajendra Prasad Central Agricultural University.**

☆ All India average consumption of fertilizer in India (2016-17) — **144 kg/ha**

☆ Soil Health Card scheme (SHC) launched in — **February 2015 from Suratgarh (Rajasthan)**

☆ Soil Health Card scheme slogan is — **"Swasth Dharaa Khet Haraa"**

☆ Paramparagat Krishi Vikas Yojna (PKVY) launched in — **2014-15 with the objective to promote organic farming**

☆ Mission Organic Value Chain Development for North Eastern Region (MOVCFNER) launched in — **11 Jaunary, 2016**

☆ Pradhan Mantri Fasal Bima Yojna (PMFBY) premium — **Maximum 2% for Kharif, 1.5% for Rabi and 5% for Annual/Horticultural crops**

☆ Pradhan Mantri Krishi Sinchayee Yojna (PMKSY) mission is — **"Per drop more crop"**

☆ AIBP, IWMP and OFWM projects merging in — **PMKSY**

☆ Bringing Green Revolution to Rastern India (BGREI) initiated in — **2010-11**

☆ National Mission on Oilseeds and Oil Palm (NMOOP) started in — **1 April, 2014**

☆ National Institute for Transforming India (NITI AYOG) set in — **1 January, 2015**

☆ NITI AYOG replaced — **Planning commision of India**

☆ Total Degraded Land — **175 mha**

☆ Severely degraded land — **65 mha**

☆ Total Area under Forest — **75 mha (19.50 per cent)**

☆ Total cropped area — **44 per cent of Geographical area**

☆ Optimum forest area required — **33 per cent of Geographical area**

☆ Maximum area under irrigation is — **Ganga basin**

☆ Maximum average annual runoff occurs in — **Brahmaputra basin**

☆ Major source of irrigation in India (2016-17) — **Tube well (57 per cent approx.) than Canal (32 per cent approx.)**

☆ Maximum irrigation area in India — **Punjab**

☆ Source of highest irrigation in Punjab — **Canal**

☆ Highest open wells found in — **Gujarat**

☆ Highest consumption of potash fertilizer — **Maharashtra**

☆ 100 per cent imported fertilizer in India — **Potash**

☆ Share of agriculture and allied sectors to the National GDP (2014-15) — **13.9**

☆ Share of livestock and fisheries to the National GDP (2014-15) — **4.5**

☆ GDP growth rate during (2016-17) — **7.1 per cent**

☆ GDP annual growth rate in India — **6.04 per cent**

☆ Contribution of agriculture and Allied on GDP during (2016) — **17.3 per cent**

☆ Annual growth rate of milk production during (2013-14) — **3.97 per cent**

☆ Total food grain production during (2016-17) — **253.2 mt**

☆ Milk production in India during (2015-16) — **155.5 mt**

☆ Milk availability g/capita/day in India during (2015-16) — **337 gms**

☆ Egg production in India during (2015-16) — **78.4 billion nos.**

☆ Egg availability nos./capita/year in India during (2014-15) — **51**

☆ Per capita land availability in India (1991-92) — **0.37 ha**

☆ Per capita arable land availability in India (according to the World Bank, 2011) — **0.13 ha**

☆ Per capita agriculture land availability in India (1991-92) — **0.16 ha**

☆ Fruit production in India (2014-15) — **88.9 Mt**

☆ Vegetable production in India (2014-15) — **162.8 Mt**

☆ Spices production in India during 2014-15 — **5.9 Mt**

☆ Flower production in India during 2014-15 — **2.29 Mt (loose)**

☆ Total horticulture produce including fruits, vegetables, flowers and spices — **280.70 Mt**

☆ National Biodiversity Board Situated at — **New Delhi**

☆ How many fat found in 100 g of chicken egg — **10 g (approx.)**

☆ Biodiversity hotspots — **28**

☆ India shares in world for water — **4.2 per cent**

☆ India shares in world for total geographical area — **2.4 per cent**

☆ India shares in world for population — **17 per cent**

☆ India shares in world for cattle population — **18 per cent**

☆ India shares in world for livestock population   **– 11 per cent**

☆ India shares in world for forest   **– 1.5 per cent**

☆ India's share of world trade   **– < 5 per cent**

☆ CACP recommended MSP for How many crops   **– 25**

☆ Types of soils according to ICAR is   **– 8**

☆ Government provided subsidy on fertilizer prices to farmers   **– 60-75 per cent**

☆ Contribution of Indian agriculture to livelihood during (2016) **– 58 per cent**

☆ Total fertilizer consumption is maximum in   **– U.P.**

☆ Highest fertilizer consumption is found in (2016-17)   **– Punjab (266 kg/ha)**

☆ State having Highest geographical area state   **– Rajasthan**

☆ State having Highest forest area   **– Madhya Pradesh**

☆ Major pulse producing state in India   **– Madhya Pradesh**

☆ Lowest forest area of India   **– Western Rajasthan**

☆ State having Highest water erosion   **– West Bengal**

☆ State having Highest wind erosion   **– Rajasthan**

☆ Highest rainfall occurs in India   **– Mousinram (Meghalaya)**

☆ Total registered pesticides in India   **– 275**

☆ Total bannel pesticides in India   **– 28**

☆ Total ATARI   **– 11**

☆ Total KBK (April, 2017   **– 680)**

☆ Lowest rainfall occurs in India   **– Jaiselmer (Rajasthan)**

☆ Fort of soybean   **– Madhya Pradesh**

☆ Garden city of India   **– Banglore**

☆ Bench terracing usually practiced on slope ranging from   **– 16-33 per cent**

☆ Export of wheat has been prohibited since   **– 8 October 2007**

☆ Export of non-basmati rice has been prohibited since   **– 15 October 2007**

☆ First agriculture census conducted in India   **– 1970**

☆ Insecticide Act was passed by GOI in   **– 1968**

☆ Pesticide restricted for use in India   **– 13**

☆ Number of Insecticide approved to control household pets   **– 39**

☆ Nutrient found in egg for brain development is   **– Choline**

- ☆ National animal of India — **Tiger**
- ☆ National bird of India — **Peacock**
- ☆ National tree of India — **Banyan**
- ☆ 1st livestock census conducted in India — **1919**
- ☆ 19th livestock census conducted in India — **2011**
- ☆ Agriculture census conducted in — **Every 5 Year**
- ☆ Livestock census conducted in — **Every 4 Year**
- ☆ Economic census conducted in — **Every 7 Year**
- ☆ National mission on micro irrigation (NMMI) has been implemented in — **2005-06**
- ☆ Year announced as the Water Year — **2007**
- ☆ Swajaldhara drinking water project is run since — **2002**
- ☆ Krishi karman award were presented for the first time in — **16th July 2011**
- ☆ E-chaupal established by — **ITC for M.P.**
- ☆ Mausam Bhawan is situated in — **New Delhi**
- ☆ Most abundant protein present in the world is — **Rubisco**
- ☆ Nursery area required for seedling of paddy for one hectare field is — **0.10 ha**
- ☆ Maximum Sale Price has been by — **Central Government (for Bt Cotton)**
- ☆ Minimum Support Price (MSP) covers 25 commodities of — **23 Crops (Cereals -7, Pulses -5, Oilseed -8, Other -5)**
- ☆ Chairman of National Commission on Farmers (NCF) — **M.S. Swaminathan**
- ☆ Directorate of Marketing Research and Inspection (DMRI) situated at — **Nagpur**
- ☆ According to ITK Bael fruit can be used to control — **Paddy blast**
- ☆ Staple food of India — **Rice**
- ☆ Staple food if Asia — **Wheat**
- ☆ National green tribunal (NGT) was established in — **October 2010**
- ☆ According to ITK Cow urine can be used to control — **Termite and Sorghum smut**
- ☆ Krishi karman award giving for — **Total food grains production**
- ☆ Three major soil groups of India — **Alluvial, Black and Red soil**
- ☆ Alluvial soil cover about — **78 Mha (24 per cent) of total land**

☆ Krishi karman award for paddy crop (2014-15) — **Chhattishgarh**

☆ Central grain analysis laboratory situated at — **Krishi Bhawan, New Delhi**

☆ Highest award presented to an agricultural scientist in the country — **Rafi Ahmad Kidwai award**

☆ Gas responsible for Bhopal gas tragedy in 1984 — **Methyl isocyanate (MIC)**

Availability of Agriculture Products/Capita/Day

Particular	Requirement (Per Capita Per Day)	Availability (Per Capita per Day) (2009-10)	Availability (Per Capita per Day) (2014-15)
Cereals	475 g	410 g	468.9 g
Pulses	80 g	29 g	41.90 g
Food grains	—	—	510.8 g
Fruits	120 g	80 g	80 g
Vegetables	275 g	240 g	240 g
Milk	200 g (non- vegetarians) 300 g (vegetarians)	245 g	307 g
Egg	182 eggs/capita/year	45 eggs/capita/year	51 eggs/capita/year

Minimum Support Price

Crop	2015-16 Price (Rs/q)	2016-17
Common Paddy	1410	1470
Grade A Paddy	1450	1510
Wheat	1525	1625
Maize	1325	1365
Barley	1225	1325
Jowar (Hybrid)	1570	1625
Gram	3425	4000
Lentil	3325	3950
Arhar	4625	5050
Moong	4850	5225
Urad	4625	5000
Cotton (Multi staple)	3800	3860
Cotton (Long staple)	4100	4160
Soybean (Black)	-	-
Soybean (Yellow)	2600	2775
Mustard	3350	3700

Crop	2015-16 Price (Rs/q)	2016-17
Sunflower	3800	3950
Jute	-	-
Safflower	3300	
Sesamum	4700	5000
Groundnut (in shell)	4030	4220
Sugarcane	-	-
Niger (seed)	3650	3825
Bajra	1275	1330
Ragi	1650	1725

Important International Years (Celebrated by UNESCO)

International year of Paddy, International Year to Commemorate the Struggle against Slavery and its Abolition	2004
International year of Micro credit, Sport and physical education, Physics	2005
International year of desert and desertification	2006
International year of polar year	2007
International year of potato, Planet earth, Languages, Sanitation	2008
International year of Natural fibres, Astronomy, Gorilla, Reconciliation and Human rights learning	2009
International year of biodiversity, Youth, Seafarer, and Rapprochement of cultures	2010
International year of People of African Descent, Chemistry, Youth and Forests	2011
International year of Cooperative and Sustainable Energy for All	2012
International year of Water cooperation and Quinoa	2013
International year of family farming, Solidarity with the Palestinian People, Small Island Developing States, Crystallography	2014
International year of Soil, light and light based technology	2015
International year of pulses	2016
International Year of Sustainable Tourism for Development	2017

Father of Different Disciplines

Father of	Name of Scientist
Agronomy	Pietrode Crescenzi
Agrometeorology	D. N. Walia
Agroclimatology	Koppen
Anatomy	Malpighi
Ayurveda	Charak
Agriculture chemistry	Justus von Liebig
Biochemistry	Justus von Liebig
Biology	Aristotle

Father of	Name of Scientist
Botany	Theophrastus
Bacteriology	Robert Koch
Bryology	Johann Hedwig
Blood group	Karl Landsteiner
Cytoplasmic Inheritance	Karl Correns
Cooperative movement in India	F. Nicholson
Extension education	Seaman A. Knapp
Experimental genetics	T. H. Morgan
Experimental physiology	Galen
Field plot experiment	J. B. Boussingault
Fruit and vegetable preservation	Nicolas Appert
Fermentation	Louis Pasteur
Genetics	G. J. Mendel
Green revolution	N. E. Borlaug
Green revolution in India	M. S. Swaminathan
Golden revolution in India	K. L. Chadha
Golden rice	Ingo Potrykus
Hybrid rice	Yuan Longping
Hybrid cotton	C. T. Patel
Indian plant pathology	E. J. Butler
Indian mycology	E. J. Butler
Indian rust	K. C. Mehta
Immunology	Edward Jenner
Indian taxonomy	Henry Santapau
Indian horticulture	M. H. Marigowda
Modern cytology	C. P. Swanson
Microbiology	Louis Pasture
Modern anatomy	A. Vesalius
Modern Genetics	G. Mendel
Mutation	Hugo de Vries
Medicine	Hippocrates
Mycology	P. A. Micheli
Modern Bacteriology	Louis Pasteur
Modern soil	Dokuchaev
Nematology	N. A. Cobb
Ornamental gardening	M. S. Randhawa
Plant pathology	Anton de Bary
Plant physiology	Stephen Hales
Pomology	De Candolle
Indian Plant physiology	J. C. Bose

Father of	Name of Scientist
Pedology	V. V. Dokuchalev
Plant tissue culture	G. Haberlandt
Sociology	Auguste Compte
Statistics	R. A. Fisher
Soil science	Vasily Dokuchaev
Soil microbiology	S. N. Winogradsky
Super rice	G. H. Khush
Soil testing	M. L. Troug
Taxonomy	Carolus Linnaeus
Zero Tillage	Jethro Tull
Weed Science	Jethro Tull
White revolution	Varghese Kurien
Zoology	Aristotle

Different Agriculture Term and Coined by

Term	Coined by
Amitosis	P. Oudet (1975)
Allosome and Heterosome	Montgomery
A chromosome	L. F. Randolph (1928)
Allelomorph	W. Bateson (1902)
Antibiotic	Selman Waksman (1942)
Biosystem	Thienemann (1939)
Biocoenosis	Karl Mobius (1877)
Bioenert body	Vernadsky (1944)
Bacteriophage	F. Herelle (1917)
B chromosome	L. F. Randolph (1928)
Chloroplastid	Schimper
Cytogenetics	Muller
Chromatin	W. Flemming (1879)
Chromosome	W. Waldeyer (1988)
Cistron	S. Benzer (1955)
Coupling phase	W. Bateson and R. C. Punnet (1906)
Coincidence	Muller
Diffusion pressure deficit	B. S. Meyer
Dosage compensation	H. J. Muller (1948)
Endoplasmic reticulum	K. R. Porter (1953)
Enzyme	W. Kuhne (1878)
Epistatic gene	W. Bateson (1902)
Eukaryote	E. Chatton (1937)

Term	Coined by
Exon	W. Gilbert (1978)
Ecosystem	A. G. Tansley (1935)
Ethology	G. S. Hilaire (1859)
F1 and F2	W. Bateson (1902)
Gene and Genotype	W. Johannsen (1909)
Genetics	W. Bateson (1905)
Hormone	Starling
Humus theory	Liebig
Hetero chromatin	E. Heitz (1928)
Heterozygote	W. Bateson (1902)
Histone	Kossel (1890)
Homozygote	W. Bateson (1902)
Intron	W. Gilbert (1978)
Intersex	R. B. Goldsmith (1915)
Interference	Muller
Integrated pest management	Geiger and Clark (1961)
Lysosome	C. Duve (1955)
Meiosis	Farmer (1905)
Mitosis	W. Flemming (1882)
Mitochondria	C. Benda (1890)
Nucleolus	Bowana
Nucleosome	Waldeyer (1889)
Nucleoplasm or Karyolymph	Strasburger (1882)
Ockologie	Ernst Haeckel (1869)
Protein	Moulder (1840)
Plastid	Haeckel
Pinocytosis or Cell Drinking	W. H. Lewis (1934)
Phagocytosis or Cell Eating	Metchnikoff
Phytohormone	Thieman
Polygene	Mather
Penicillin	Alexander Fleming (1939)
Root pressure	Renner
Ribosome	Palade (1955)
Suction pressure	Renner
Synecology	Schroter and Kirchner (1902)
Stimulus	Darwin
Thylakoids	Menke (1962)
Vernalization	Lysenko
Vitamin	Funk

Term	Coined by
Virus	D.J. Ivanowsky (1872)
Viroids	T.O. Diener (1871)
Weed Science	Jethro Tull
Zero tillage	Jethro Tull

Causes of Pungency and Colour in different Crops

Bitterness in almond	Imailedin
Bitterness in pepper	Marmelosin
Odour in pepper	Oleriosin
Bitterness in cucumber	Cucurbitacin
Bitterness in bael	Marmelocin
Pungency in radish	Isothiocynite
Pungency in chilli	Capsicin alkaloid
Flavour in onion	Allyle propyl di-sulphide
Pungency in crushed garlic	Allycin
Pungency in uninjured garlic	Allinase, Amino acid
Flavour in bittergourd	Memordicoside, Tetra tri-terpine
Flavour inaonla	Tennin/Pollophenols
Acidity in arbi	Calcium oxalate
Acidic taste in gram leave	Malic acid/Oxalic acid/Tenin
Pungency of toria	Allyl iso-thiocynate
Pungency of mustard oil	Glycoside singrin
Bitterness in aonla	Polyphenole and tenin
Red colour in tomato	Lycopene
Yellow colour in papaya	Caricaxanthin pigment
Yellow colour in onion	Quercetin
Yellow colour in turmeric	Curcumin
Orange colour in carrot	Carotene
Red colour in carrot	Anthocyanin
Red colour in chiili	Capsanthin
Green colour in potato	Solanin

Superlatives of Plants

Largest flower	*Rafflesia arnoldi*
Largest fruit	*Lodoicea maldivea*
Smallest seed	Orchids
Smallest flower	*Wolffia*
Lightest wood	*Ochroma pyramidale*

Smallest bacteria	*Dialister pneumosintes*
Largest ovule	*Cycas circinalis*
Largest seed	*Lodoicea seychelles*
Largest leaf	*Victoria regia* (angiosperm)
Largest inflorescence	*Puya raimondii*
Smallest Mycoplasma	*Mycoplasma gallisepticum*
Largest unicellular organism	*Acetabularia* (marine green algae)
Largest fern	*Alsophila excelsa*
Tallest tree	*Sequoia gigantea* (Gymnosperm)
Tallest angiosperm -	Eucalyptus
Largest herb	*Musa paradisiaca*
Largest grass	Bamboosa
Smallest pteridophyte	*Azolla*
Smallest gymnosperm	*Zamia pygmaea*
Smallest angiosperm	*Wolffia microscopica*
Minimum number of chromosome	*Haplopappus gracilis (2n=4)*

Classification of Crops According to Family

Family Name	*Crop Name*
Gramineae (Poaceae)	All cereals, Oat, Sugarcane, Napier grass
Leguminosae (Papilionaceae)	Pulses, Groundnut, Sunhemp, Berseem, Guar, Lucerne
Cruciferae (Brassicaceae)	Mustard, Radish, Cabbage, Cauliflower
Malvaceae	Cotton, Okra
Tiliaceae	Jute
Solanaceae	Tobacco, Potato, Brinjal, Tomato, Chilli
Cucurbitaceae	All gourds, Pumpkin, Cucumber, All melon
Compositae (Asteraceae)	Sunflower, Safflower, Niger
Pedaliaceae	Sesamum
Euphorbiaceae	Castor, Topica
Linaceae	Linseed
Chenopodiaceae	Sugarbeet, Spinach
Convolvulaceae	Sweet potato
Umbelliferae	Coriander, Cumin, Carrot
Aliaceae	Onion, Garlic
Zingiberaceae	Turmeric, Ginger
Rutaceae	Citrus, Bael
Musaceae	Banana
Anacardiaceae	Mango
Rosaceae	Apple
Myrtaceae	Black plum, Guava

Botanical Name and Family of Agronomical Crops

Crops	Botanical Name	Family
Cereals crops		
Paddy	*Oryza sativa*	Gramineae
Wheat	*Triticum aestivum*	Gramineae
Barley	*Hordeum vulgare*	Gramineae
Maize	*Zea mays*	Gramineae
Bajra (Pearl millet)	*Pennisetum glaucum*	Gramineae
Jowar (Sorghum)	*Sorghum vulgare*	Gramineae
Triticale	*Secale cereale*	Gramineae
Buck wheat/Pseudo cereal	*Fagopyrum esculentum*	Gramineae
Millet crops		
Cheena (Proso millet)	*Panicum miliaceum*	Gramineae
Kakun (Foxtail/Jerman millet)	*Setaria italica*	Gramineae
Kodo (Coarsest millet)	*Paspalum scrobiculatum*	Gramineae
Ragi (Finger millet/Madua)	*Eleusine coracana*	Gramineae
Sawan (Barnyard millet)	*Echinochloa frumentancea*	Gramineae
Pulse crops		
Gram (Chick pea/Bengal gram)	*Cicer arietinum*	Leguminosae
Field pea	*Pisum sativum* var. *arvense*	Leguminosae
Garden pea	*Pisum sativum* var. *hartense*	Leguminosae
Arhar (Red gram/Pigeon pea)	*Cajanus cajan*	Leguminosae
Soybean	*Glycine max*	Leguminosae
Urad (Black gram)	*Phaseolus mungo*	Leguminosae
Moong (Green gram)	*Phaseolus aureus*	Leguminosae
Rajma (French bean)	*Phaseolus vulgaris*	Leguminosae
Lentil	*Lens esculenta*	Leguminosae
Cow pea (Lobia)	*Vigna unguiculata*	Leguminosae
Grass pea (Lathyrus)	*Lathyrus sativus*	Leguminosae
Moth bean	*Phaseolus aconotifolia*	Leguminosae
Kulthi (Horse gram)	*Macrotyloma uniflorum*	Leguminosae
Edible oilseed crops		
Ground nut (Peanut/Monkey nut)	*Arachis hypogea*	Leguminosae
Til (Sesamum)	*Sesamum indicum*	Pedaliaceae
Sunflower	*Helianthus annus*	Compositae
Safflower	*Carthamus tinctorius*	compositae
Mustard and Rapeseed	*Brassica* spp.	Crucifereae
Niger	*Guzotta abssicinia*	

Crops	Botanical Name	Family
Non edible oilseed crops		
Linseed (Flex)	*Linnum usitatissimum*	Linaceae
Castor	*Ricinus communis*	Euphorbiaceae
Fiber crops		
Cotton	*Gossypium* spp.	Malvaceae
Jute	*Corchorus capsularis*	Tiliaceae
Sunhemp	*Crotolaria juncea*	Leguminosae
Forage crops		
Oat	*Avena sativa*	Gramineae
Gaur (Cluster bean)	*Cyamoposis teragonoloba*	Leguminosae
Berseem	*Trifolium alexandrinum*	Leguminosae
Lucerne (Alfalfa)	*Medicago sativa*	Leguminosae
Napier grass	*Pennisetum purpureum*	Gramineae
Tuber crops		
Potato	*Solanum tuberosum*	Solanaceae
Tapioca (Cassava)	*Manihot utilissima*	
Sugar crops		
Sugarcane	*Saccharum officinarum*	Gramineae
Sugar beet	*Beta vulgaris*	Chenopodiaceae
Stimulate crops		
Tobacco	*Nicotiana* spp.	Solanaceae
Betel	*Piper betel*	Piperaceae

Origin of Field Crops

Origin	Crops Name
Indo-Burma/South East Asia	Paddy
Maxico	Maize
South East Asia	Sugarcane
South West Asia	Wheat, Barley, Gram, Lucerne, Buckwheat
India	Kodo, Kutki, Oat, Urad, Moong, Arhar, Cotton, Jute
China	Soybean, Mustard, Tea
Mexico and USA (Central America)	Tobacco, Sunflower
Brazil	Groundnut
Peru (South America)	Tomato, Potato
Afghanistan	Linseed
Egypt	Berseem
Africa	Til, Jowar, Bajra, Castor, Cowpea

Oil Percentage in different Crops

Sl.No.	Crop	Oil (per cent)	Sl.No.	Crop	Oil (per cent)
1.	Soybean	20	6.	Mustard	33
2.	Sesamum	46-52	7.	Niger	37-43
3.	Groundnut	44-50	8.	Coconut	60
4.	Castor	35-58	9.	Cotton	15-25
5.	Safflower	24-36	-	-	-

Protein Percentage in different Crops

Sl.No	Crop	Protein (per cent)	Sl.No	Crop	Protein (per cent)
1.	Safflower	40-45	9.	Cow pea	23.4
2.	Soybean	42	10.	Pea	22-23
3.	Linseed	36	11.	Sesame	18-20
4.	Groundnut	26	12.	Gram	21.1
5.	Arhar	21-25	13.	Wheat	11-12
6.	Moong	25	14.	Bajra	11-12
7.	Lentil	25	15.	Maize	10
8.	Urad	24	16.	Paddy	6-7

Term and Associated Crops

Term	Associated Crop
Arrowing	Sugarcane
Ginning	Cotton
Curing	Tobacco, Tea, Mustard
Retting, Stripping	Jute, Sunn hemp
Earthing up	Potato, Sugarcane
Ratooning	Sugarcane
Wrapping	Sugarcane
Trashing	Sugarcane
De-suckering	Tobacco
Topping, Nipping	Cotton
Nipping	Gram
De-tasseling	Maize
Depog, Medagaskar	Paddy
Proboiling	Paddy
Propping	Banana
Shelling	Groundnut
Seasoning	Turmeric, Chilli

Important Varieties of Field Crops

Crop Name	Varieties
Paddy	IR-36, IR-64, Purnima, Kranti, Mahamaya, Bumleshwari, Pusa basmati-1, Safri-17, Jagannath, GEB-24, Type-3, Madhukar, Jaya, IR-8, Lunishree, Yamini, Krishna Hansa, Vasumati, Pusa sugandha-2, Pusa-1121,
Wheat	WH-147, Sujata, GW-173, GW-273, HW-2004, Kanchan, LOK-1, C-306, Lermorojo, Sonora-64, Kalyan sona, Sonalika, HI-1077, HD-2733, HD-2824 Poorva, Urja, Pusa vishesh, Pusa gold, Aditya, HD-2888, Tripti, PBW-65,
Maize	Ganga-1, Ganga safed-2, Ranjit, Himalaya, VL-54, Navjot, Pusa composite-3, PEHM-3
Barley	NBL-11 (Sindhu), VL-56, NDB-1173
Arhar	Prabhat, UPAS-120, Type-21, Pragati, Gwalior-3, Rajiv lochan, Pusa ageti, Mukta, Amar, Azad, ICPH-8, Paras, Malviya chamatkar, Pusa 991, Pusa-9712, Bahar, ICPL-87 (Asha)
Gram	JG-315, JG-322, Radhey, Pusa-417, Vikas, Vishwas, JG-11, JG-74, Vijay, Shweta, JGK-1, Vishal, Pusa-1088, Pusa-1105, BGH-547, BG-1105
Pea	Arkel, Rachana, Paras, JP-855, Vikash, Ambika, Aparna, KPMR-144-1, Pusa pragati, Pusa mukta, Pusa panna, Telephone, New line perfection
Soybean	JS-2, JS-335, JS-93-05, PK-472, Gourav, Ankur, Durga, Pusa-9712
Groundnut	Vikram, ICGS-1, SB-11, JL-24, Junagarh-11,
Moong	Samrat, Pragya, Pusa baisakhi, JM-721, K-851, PDM-1, Pusa-105, Pusa bold, Pusa ratna, PDM-54
Urad	Pant U-30, JU-2, Type-9, Sarla, Barkha, Prabha, CO-1, Gwalior-2, UG-218
Lentil	K-75, JL-3, Lens-4076, Nuri, Shivalik, Malika, Pusa-6
French bean	Arka ambara, Arka Sharath, Arka Suvidha, Arka komal
Mustard	Kranti, Pusa bold, Vardan, Pusa kalyani, Varuna, Pusa jai kisan, Kalyani, Pusa swarnima, Pusa agrani, Pusa karishma, Mahak, Urvashi, Basanti
Til	Selection-5, JT-21, TC-25
Sunflower	Jwalamukhi, MSFH-8, KBSH-1,
Safflower	Annagiri, JSH-129, JSF-1, JS-17
Cotton	Anjali, Khandwa-2, Pratima, JKH-2, Sujata, Pusa-343, Pusa17-52-10
Sugarcane	CO-86032, CO-62175, CO-671, Rajbhog
Ramtil	IGP-76, GP-10, JNS-1
Lathyrus	Ratan, Pratik,
Castor	Kranti, Jwala, Jyoti, DCH-32, JCH-4
Linseed	Indravati (RLC-92), Kiran, Jawahar-552, Deepika,
Kodo	GPUK-3, IPS-147-1, JK-1, ICCK-737
Kutki	PRc-3, IGL-4
Kulthi	AK-21, JND-2
Ragi	VL-147, PR-202, HR-374
Moth bean	RMO-40
Pearl millet	Pusa composite-383
Cow pea	V-578, Arka garima,
Cluster bean	Goma manjari

Water Requirement of different Crops

Crop Name	Water Requirement (mm)	Crop Name	Water Requirement (mm)
Wheat	450-650	Tobacco	400-600
Paddy	900-2500	Tomato	600-800
Maize	500-800	Pea	350-500
Sorghum	450-650	Onion	350-550
Sugarcane	1500-2500	Cabbage	350-500
Sugar beet	550-750	Banana	1200-2200
Soybean	450-700	Citrus	900-1200
Groundnut	500-700	Grape	500-1200
Cotton	700-1300	Pineapple	700-1000
Sugarcane	1500-2500		
Pearl millet	450-600		

Critical Stages of Irrigation

Crop Name	Critical stages
Wheat	CRI, shooting and heading
Paddy	Panicle initiation, heading and flowering
Maize	Tasseling, silking and early grain formation
Sorghum	Booting, flowering, milky and dough stage
Sugarcane	Formative stage and tillering
Barley	Shooting and heading
Soybean	Flowering and seed formation
Groundnut	Flowering, peg penetration and early pod development
Cotton	Flowering and boll development
Finger millet	Panicle initiation and flowering
Pearl millet	Heading and flowering
Sesame	Flowering
Sunflower	Flower bud initiation, head initiation, flowering and milky stage
Safflower	Flower bud initiation
Chilli	Flowering
Potato	Tuber initiation to tuber maturity
Onion	Bulb formation to maturity
Pea	Flowering and pod development
Gram	Flowering and pod development
Pigeon pea	Flower bud initiation and pod development

Test Weight of different Crops

Sl.No.	Crop	Test Weight/Seed Index (g)
A. Cereal (1000-grain weight)		
1.	Paddy	20-30
2.	wheat	36-45
3.	Barley	35-40
4.	Maize	200-250
5.	sorghum	25-30
6.	Pearl millet	5-9
B. Pulses (100-grain weight)		
1.	Chick pea	9.6 (small), 18.5 (medium), 41 (large)
2.	Pigeon pea	4.5-10.5
3.	Lentil	1.7-3.8
4.	Black gram	3.5-5.0
5.	Green gram	2-7
6.	pea	12-22
Sl.No.	Crop	Test Weight/Seed Index (g)
C. Oilseeds		
1.	Mustard	3-5
2.	Groundnut	20-25
3.	Sunflower	40-50
4.	Safflower	50-60
5.	Castor	10-15
D. Others		
1.	Cotton	7-10
2.	Lucern	2-4

☆ Test weight - weight of 1000 seeds. (used for small seeded)

☆ Seed Index – weight of 100 seeds. (used for bold seeded)

Fruit Types and Edible Part of Agronomical Crops

Crop Name	Fruit Type	Edible Part
All cereals and grasses	Caryopsis	Endosperm and Embryo
Gram, Pea, Arhar	Pod/Legume	Cotyledons/Seed
Ground nut	Lomentum	Cotyledons/Seed

Botanical Name and Family of Important Weeds

English Name	Common Name	Botanical Name	Family
Wild oat	जंगली जई	*Avena fatua*	Gramineae
Mexican prickly poppy	सत्यानाशी	*Argemone mexicana*	Papaveraceae
Spiny amaranthus	कॉटेदार चौलाई	*Amaranthus spinosus*	Amarathceae
Bill goat weed	फुलनी / महकुआ	*Ageratum conyzoides*	Compositae
Prickly chafflower	लटजीरा / चिरचिटा	*Achyranthes aspera*	Amaranthceae
Slender amaranthus	जंगली चौलाई	*Amaranthus viridis*	Amaranthceae
Pimpemel	कृष्ण नील	*Anagaltis arvensis*	Primulaceae
Wild mustard	जंगली सरसो	*Brassica sinensis*	Cruciferae
Spreading hog weed	विष खपरा	*Boerhavia diffusa*	Nyctaginaceae
Bermuda grass	दूब घास	*Cynodon dactylon*	Gramineae
Purple nutsedge	गोथा	*Cyperus rotundus*	Cyperaceae
Yellow nutsedge/rice flat sedge	मोथा	*Cyperus iria*	Cyperaceae
Umbrella sedge	मोथा	*Cyperus difformis*	Cyperaceae
Bind weed	हिरन खुरी	*Convolvulus arvensis*	Convolvululaceae
Lambsquate/Dog tooth grass	बथुआ	*Chenopodium album*	Chenopodiaceae
Chicory/Blue daisy	कासनी	*Cichorium intybus*	Compositae
Giant swallow wort	मदार	*Calotropis gigantea*	Ascletiabaceae
Wild safflower	जंगली कुसुम	*Carthamus oxyacantha*	Compositae
Wild jute	चेज	*Corchorus acutangulus*	Tiliaceae
Dodder	अमरबेल	*Cuscuta* spp.	Convolvulaceae
White cock`s comb	सिलयारी	*Celosia argentea*	Amaranthceae
Buffalo gram	चरोटा	*Cassia tora*	Leguminosae
Jimson weed	धतुरा	*Datura stramonium*	Solanaceae
Thorne apple	कॉटेदार धतुरा	*Datura alba*	Solanaceae
Marvel grass	कॉदी	*Dicanthum annulatum*	Gramineae
Crab grass	घुड़ दूब	*Digitaria sanguinalis*	Gramninae
Pill pod spurge	छोटी दूधी	*Euphorbia hirta*	Compositae
Garden spurge	बड़ी दूधी	*Euphorbia geniculata*	Compositae
Water hyacinth	जलकुंभी	*Eichhonia crassipes*	Pontederiaceae
Jungle rice	सॉवा	*Echinochloa colonum*	Gramineae

English Name	Common Name	Botanical Name	Family
Barnyard grass	मुछवाला सावाँ	*Echinochloa crus* galli	Gramineae
False daisy	भृंगराज	*Eclipta alba*	Compositae
Goose grass	जंगली रागी	*Eleusine indica*	Gramineae
Fumitory	वन सोया / गाजरी	*Fumaria parviflora*	Fumariaceae
Swamp morning glory	जलकर्मी	*Ipomea repens*	Convolvulaceae
Wrinkle grass	टोरा टोरी	*Ischaemum rugosum*	Gramineae
Prickly lantana	जरायन	*Lantana camara*	Verbenaceae
Lathyrus	खेसरी	*Lathyrus sativus*	Leguminoceae
Touch me not	लाजवंती	*Mimosa pudica*	Leguminoceae
Touch me not	कॉटेदार लाजवंती	*Mimosa spinosa*	Leguminoceae
White sweet clover	सफेद सेंजी	*Melilotus alba*	Leguminoceae
Yellow sweet clover	पीली सेंजी	*Melilotus indica*	Leguminoceae
Prickly pear	नागफनी	*Opuntia dilenaii*	Cacaceae
Wild rice	जंगली धान	*Oryza sativa* var. *fatua*	Gramineae
Sorrel	खट्टी बुटी	*Oxalis acetosella*	Oxalidaceae
Indian sorrel	खट्टी बुटी	*Oxalis acetosella*	Oxalidaceae
Haory basin	बन तुलसी	*Ocimum canum*	Labiatae
Broom rape	बिल्ली	*Orobanche* spp.	Orobanchaceae
Canary grass	गेहूँ का मामा	*Phalaris minor*	Gramineae
Corn spurry/Niruri	हजार दाना	*Phyllanthus niruri*	Euphorbiaceae
Purslane	जंगली पालक	*Portulaca oleracea*	Portulaceae
Purslane	नुनिया	*Portulaca quadrifolia*	Portulaceae
Knot grass	—	*Paspalum sanguinale*	Gramineae
Congress grass/Wild carrot grass	गाजर घास	*Parthenium hysterophorus*	Compositae
Ground cherry/Hog weed	चिरपोटी	*Physalis minima*	Solanaceae
Com spurry	सतगठिया / वनधनिया	*Spergula arvensis*	Caryophyllaceae
Tiger grass	कॉस	*Saccharum spontaneum*	Gramineae
Black night shade	मकोई	*Solanum nigrum*	Solanaceae
Yellow Sida	बरयारा	*Sida rhombifolia*	Malvaceae
Green Sida	बरयारा	*Sida spinosa*	Malvaceae
Witch weed	अगिया	*Striga lutea*	Scrophulariaceae

English Name	Common Name	Botanical Name	Family
Prickly brinjal	भट कटैया	*Solanum xanthocarpum*	Solanaceae
Johnson grass	बरू	*Sorghum halepanse*	Gramineae
Green fox tail	बंदरा बंदरी	*Sataria glauca*	Gramineae
Wild mint	जंगली अकरकरा	*Spilanthus comelia*	Compositae
Cattail	टायरा	*Typha* spp.	Typhaceae
Maxican daisy	बारहमासी	*Tridex procumbens*	Compositae
Carpet weed	पथरचटा	*Trianthema monegyna*	—
Vetches	टकरा अकरी	*Vicia sativa*	Leguminoceae
Common vetche	मुनमुना	*Vicia hirsute*	Leguminoceae
Cocklebur/Bur- weed	बड़ी गोखरू	*Xanthium strumerium*	Compositae
Wild ber	झरबेरी	*Zizyphus rotundifolis*	Rhamnaceae

Botanical Name and Family of Horticultural Crops

English Name	Common Name	Botanical Name	Family
Fruit Crops			
Banana	केला	*Musa paradisiaca*	Musaceae
Mango	आम	*Mangifera indica*	Anacardiaceae
Papaya	पपीता	*Carica papaya*	Caricaceae
Apple	सेब	*Malus domestica*	Rosaceae
Aonla	ऑवला	*Emblica officinalis*	Euphorbiaceae
Karonda	करौंदा	*Carissa carandus*	Apocynaceae
Guava	अमरुद	*Psidium guajava*	Myrtaceae
Java plum/Black plum/Indian black berry	जामुन	*Syzygium cumini*	Myrtaceae
Rose apple/Malabar plum/ Jambos	गुलाब जामुन	*Eugenia jambolana/ Syzygium jambos*	Myrtaceae
Indian plum/Indian jujube/ Chinese date/Chinse apple	बेर	*Zizyphus mauritiana*	Rhamnaceae
Grape	अंगूर	*Vitis vinifera*	Vitaceae
Bael	बेल	*Aegle marmelos*	Rutaceae
Lime	कागजी नींबू	*Citrus aurantifolium*	Rutaceae
Kinnow	—	*Citrus deliciosa*	Rutaceae

English Name	Common Name	Botanical Name	Family
Lemon	नींबू	*Citrus limon*	Rutaceae
Orange	संतरा	*Citrus reticulata*	Rutaceae
Sweet orange	मौसंबी	*Citrus sinensis*	Rutaceae
Chakotara	चकोतरा	*Citrus maxima*	Rutaceae
Coconut	नरियल	*Cocos nucifera*	Aricaceae
Fig	अंजीर	*Ficus carica*	Moraceae
Strawberry	रसभरी	*Fragaria* spp.	Rosaceae
Phalsa	फालसा	*Grewia subinaequalis (old – Grewia asiatica)*	Tiliaceae
Cashewnut	काजू	*Anacardium occidentale*	Anacardiaceae
Walnut	अखरोट	*Juglans regia*	Fagaceae
Litchi	लीची	*Litchi chinensis*	Sapindaceae
Date plam	खजूर	*Phoenix dactylifera*	Palmae (Arecaceae)
Almond	बादाम	*Prunus armeniaca*	Rosaceae
Apricot	खुबानी	*Prunus armeniaca*	Rosaceae
Pear	नाशपाती	*Prunus communis*	Rosaceae
Plum	आलू बुखारा	*Prunus domestica*	Rosaceae
Peach	आड़ू	*Prunus persica*	Rosaceae
Pecanut/Pecan	पिस्ता	*Carya illinoensis*	Juglandaceae
Pomegranate	अनार	*Punica granatum*	Punicaceae
Raspberry	रसबेर	*Rubus idaeus*	Rosaceae
Cape goose berry	—	*Physalis peruviana/ Grewia asiatica*	Rosaceae
Tamarind	इमली	*Tamarindus indica*	Fabaceae/ Leguminaceae
Pine apple	अनानास	*Annanas comosus*	Bromeliaceae
Wood apple	कबिट	*Feronia limonia/Limonia acidissima*	Rutaceae
Ramphal	शरीफा	*Annona reticulata*	Annonaceae
Custard apple	सीताफल	*Annona squamosa*	Annonaceae
Kiwi fruit	—	*Actinidia chinensis*	Actinidiaceae
Mulberry	शहतूत	*Morus alba*	Moraceae
Sapodilla	चीकू	*Achras sapota*	Sapotaceae

English Name	Common Name	Botanical Name	Family
Monster plant	अमर फल	*Monstera deliciosa*	Aricaceae
Loquat	—	*Ertobotrya japonica*	Rosaceae
Areca nut/Betel nut	सुपाडी	*Areca catechu*	Aricaceae
Water chestnut	सिघाड़ा	*Trapa bispinosa*	Trapaceae
Vegetable Crops			
Tomato	टमाटर	*Solanum lycopersicon*	Solanaceae
Brinjal	बैंगन	*Solanum melongena*	Solanaceae
Chilli	मिर्च	*Capsicum annum*	Solanaceae
Carrot	गाजर	*Daucus carota*	Apiaceae
Coriander	धनिया	*Coriandrum sativum*	Apiaceae
Garden pea	मटर	*Pisum sativum*	Papilionaceae
Potato	आलू	*Solanum tuberosum*	Solanaceae
Okra	भिंडी	*Abelmoschus esculantus*	Malvaceae
Onion	प्याज	*Allium cepa*	Liliaceae
Garlic	लहसुन	*Allium sativum*	Liliaceae
Cabbage	पत्ता गोभी	*Brassica oleracea* var. *capitata*	Brassicaceae
Cauliflower	फूल गोभी	*Brassica oleracea* var. *botrytis*	Brassicaceae
Brussel`s broccoli	—	*Brassica oleracea* var. *gemmifera*	Brassicaceae
Knol-khol	गाठ गोभी	*Brassica caulorapa*	Brassicaceae
Turnip	शलजम	*Brassica rapa*	Brassicaceae
Raddish	मूली	*Raphanus sativus*	Brassicaceae
Cucumber	खीरा	*Cucumis sativus*	Cucurbitaceae
Musk melon	खरबूजा	*Cucumis melo*	Cucurbitaceae
Snap melon	—	*Cucumis melo* var. *momordica*	Cucurbitaceae
Long melon/Snake cucumber	ककड़ी	*Cucumis melo* var. *utillisium*	Cucurbitaceae
Gherkin	—	*Cucumis anguria*/ *C.sativus*	Cucurbitaceae
Water melon	तरबूजा	*Citrullus latanus*	Cucurbitaceae
Round melon	—	*Citrullus latanus* var. *fistulosus*	—

English Name	Common Name	Botanical Name	Family
Pumpkin	कुम्हड़ा / कद्दू	*Cucurbita moschate/ Cucurbita pepo*	Cucurbitaceae
Bottle gourd	लौकी	*Lagenaria siceraria*	Cucurbitaceae
Ridge gourd	तुरई	*Luffa acutangula*	Cucurbitaceae
Round gourd	टिण्डा	*Cotrullus vulgaris* var. *fistulosus*	Cucurbitaceae
Sponge gourd	रेरुआ	*Luffa cylindrica*	Cucurbitaceae
Pointed gourd	परवल	*Trichosanthus dioca*	Cucurbitaceae
Snake gourd	चिचिन्डा	*Trichosanthus anguina*	Cucurbitaceae
Ash gourd/Wax gourd	पेठा	*Benincasa hispida*	Cucurbitaceae
Ivy gourd	टिण्डौरी	*Coccinia indica/Coccinia grandis*	Cucurbitaceae
Spine gourd	—	*Momordica chinensis/ M. dioica*	Cucurbitaceae
Bitter gourd	करेला	*Momordica charantia*	Cucurbitaceae
Celery	अजमूद	*Apium graveolens*	Apiaceae
Elephant foot yam	जंगली सुरन	*Amorphophyllus campanulatus/A. paeonifolius*	Araceae
Asparagus	शतावर	*Amorphophyllus officinalis*	Asparagaceae
Beet root	चुकन्दर	*Beeta vulgaris*	Chenopodiaceae
Palak/Indian spinach/ Spinach beet	—	*Beeta vulgaris* var. *bengalensis*	Chenopodiaceae
Spinach	पालक	*Spinacea oleraceae*	Chenopodiaceae
Sweet potato	शकरकंद	*Ipomea batatas*	Convolvulaceae
Sweet pepper	शिमला मिर्च	*Capsicum annum*	Solanaceae
Fenugreek	मेथी	*Trigonella foenugraecum*	Papilionaceae/ Fabaceae
Cow pea	—	*Vigna unguiculate*	Leguminosae/ Fabaceae
Cluster bean	ग्वार	*Cymopsis tetragonalabus*	Papilionaceae
French bean	सेम	*Phaseolus vulgaris*	Leguminosae
Jack fruit	कटहल	*Autocarpus heterophyllus*	Moraceae
Lablab bean	सेम	*Lablab purpureus*	Papilionaceae
Taro	अरबी	*Colocasia antiquorum*	Araceae
Lettuce	सलाद	*Lectuca sativa*	Asteraceae
Goose foot	बथुआ	*Chenopodium album*	Chenopodiaceae

English Name	Common Name	Botanical Name	Family
Chaulai	चौलाई	*Amaranthus viridis*	Amaranthaceae
Flower crops			
Rose	गुलाब	*Rosa indica*	Rosaceae
Marigold	गेंदा	*Tagetes* spp.	Calenduleae
Chrysanthemum	गुलदाउदी	*Chrysanthemum* spp.	Asteraceae
Gladiolus/Sword lily	—	*Gladiolus spp./G. daleni/G. natalensis*	Iridaceae
Carnation	गुलनार	*Dianthus spp./D. caryophyllus*	Caryophyllaceae
Bougainvillea	—	*Bougainvillea* spp.	Nyctaginaceae
Jasmine	चमेली	*Jasminum grandiflorum*	Oilaceae
Dahlia	—	*Dahlia pinnata*	Asteraceae
Tuberose	रजनी गन्धा	*Polianthes tuberose*	Asparagaceae
Lavender	—	*Lavandula angustifolia/L. officinalis*	Lamiaceae
Arabian jasmine	मोगरा	*Jasminum sambac*	Oliaceae
Serewpine	केवड़ा	*Pandanus odoratissimum*	Pandanaceae
Lotus	कमल	*Nelumbium nucifera*	Nelumbonaceae
Aromatic Crops			
Lemon grass	—	*Cymbopogan flexuasus/ C. citratus*	Poaceae
Mint/Mentha	पुदीना	*Menthe arvensis/mentha longifolia*	Labiatae
Khus/Vetivar	—	*Vetiveria zizanoides*	Poaceae
Basil	तुलसी	*Ocimum sanctum/ O. tenuliflorum*	Lamiaceae
Citronella	—	*Cymbopogan winterianus*	Cardiopteridaceae
Medicinal crops			
Asparagus	सफेद मुसली	*Chlorophytum borivilianum/Asparagus adscendence*	Liliaceae
Ashwagandha/winter cherry	अश्वगंधा	*Withania somnifera*	Solanaceae
Serpent wood/Rouvolfia	सर्पगंधा	*Rouvolfia serpentiana*	Apocyanaceae
Blond psyllium	इसब गुल	*Plantago ovata*	Plantaginaceae
Astatic pennywort	ब्राम्ही	*Bacopa morriei/Centella asistica*	Apiaceae
Butch	—	*Acorus calamus*	—
Nux vomica	—	*Strychnos nuxvomica*	—
Rudraksha	—	*Elaeocarpus sphaericus*	—

English Name	Common Name	Botanical Name	Family
Arjun	अर्जुन	*Terminalia arjuna*	Combretaceae
Qunine	कुनैन	*Cinchona officinale*	Rubiaceae
Sacred basil	तुलसी	*Ocimum sanctum*	Lamiaceae
Margosa	नीम	*Azadirachta indica*	Meliaceae
Aloe Vera	ग्वार पाठा	*Aloe barbadense*	Liliaceae
Bhangra	भृंगराज	*Eclipta prostrata*	Asteraceae
Turmeric	हल्दी	*Curcuma longa*	Zingiberaceae
Liquorice	मुलैठी	*Glycorrhiza glabra*	Fabaceae
Opium	अफीम	*Papaver somniferum*	Papaveraceae
Amrai	अजवाइन	*Trachyspermum ammi*	Apiaceae
Croton	जमालगोटा	*Croton tiglium*	Euphorbiaceae
Hemp	भॉंग	*Cannabis sativa*	Cannbinaceae
Guar gum	ग्वार	*Cyamopsis tetragonoloba*	Fabaceae
Bengal kino	कमरकस	*Butea monospermal*	Fabaceae
Spices crops			
Bengal cardamom	बड़ी इलायची	*Ammomum aromaticum*	Zingiberaceae
Black pepper	काली मिर्च	*Piper nigrum*	Piperaceae
Ammi	अजवाइन	*Trachyspermum ammi*	Apiaceae
Fennem	सौंफ	*Foeniculum vulgare*	Apiaceae
Cumin	जीरा	*Cuminum cymimum*	Apiaceae
Tamarind	इमली	*Tamarindus indica*	Caesalpinaceae
Cardomon	इलायची	*Elettaria cardamomum*	Zingiberaceae
Pepper mint	पोदीना	*Mentha piperata*	Lamiaceae
Indian cassia	तेजपात	*Cinamomum tamala*	Lauraceae
Curry leaf	मीठी नीम	*Murraya koenigii*	Rutaceae
Zinger	अदरक	*Zingiber officinale*	Zingiberaceae
Asaloetida	हींग	*Ferula asafoetida*	Apiaceae
Cinnamon	दाल चीनी	*Cinnamomum zelyanicum*	Lauraceae
clove	लौंग	*Syzgium aromaticum*	Myrataceae
Saffron	केसर	*Crocus sativus*	Eridaceae
Forest trees			
Sissoo	शीशम	*Delbergia sissoo*	Papilionaceae

English Name	Common Name	Botanical Name	Family
Sal	साल	*Shorea robusta*	Dipterocarpaceae
Teak	सागौन	*Tectona grandia*	Verbenaceae
Acacia babool	बबूल	*Acacia nilotica*	Mimosaceae
Burflower tree	कदम	*Anthocephalus indicus/ Neolamarckia cadamba*	Rubiaceae
Bamboo	बांस	*Bambusa arundinacea*	Poaceae
Orchid tree	कचनार	*Bauhinia variegata*	Caesalpiniceae
	पलाश	*Butea monospermal*	Fabaceae
Golden rain tree	अमलताश	*Cassia fistula*	Fabaceae
Flame tree	गुलमोहर	*Delonix regia*	Caesalpiniceae
Eucalyptus	नीलगीरी	*Eucalyptus tereticornis/E. globulus*	Myrtaceae
Banyan	बरगद	*Ficus banghalensis*	Moraceae
Bodhi tree	पीपल	*Ficus religiosa*	Moraceae
Tasminian blue gum	सुबबूल	*Leucaena leucocephala*	Myrtaceae
Mahua	महुआ	*Madhuca latifolia*	Sapotaceae
Karanja	कंरज	*Pongamia pinnata/ Millettia pinnata*	Fabaceae
Poplar	पॉपुलर (चिनार)	*Populus deltoides*	Salicaceae
Malabar kino	बीजा	*Pterocarpus marsupium*	Fabaceae
	अर्जुन	*Terminalia arnuna*	Combretaceae
Chebulic myrobalan	हर्रा	*Terminalia chaibula*	Fabaceae

Important Varieties of Horticultural Crops

Crop Name	Varieties
Mango	Alphonso, Arka aruna, Ratna, Amrapali, Sindhu, H-39, Pusa arunima, Pusa Surya, Mallika, Arka anmol, Keshar, Rumani, Arka Puneet
Grape	Arka neelmani, Dilkush, Arka soma, Arka Trashna, Thompson, Arka shweta, Pusa urvashi, Pusa navrang, Beauty seedless, Pusa seedless, Arka chitra, Arka majestic, Arka shyam, Arka trishna, Shweta seedless, Dogridge
Citrus	Vikram, Pramalini, Flying dragon
Apple	Winter banana, Scartlet gol, Red fuzi
Banana	FHIA-01, FHIA-03, Panchananda, Bluggoe, Saba
Strawberry	Pusa early dwarf, Pajaro
Ber	Goma kirti
Avocado	Furete

Crop Name	Varieties
Papaya	Pusa nanha, Pusa delicious, Pusa majestry, Surya, Pusa gaint, Arka probhath, Coorg honey dew, Sunehari, Lal ambari, Surya
Guava	Lalit, Safeda, Sardar (Lucknows-49), Pusa srijnar, Pusa vamankar, Pusa saummya, Arka amulya, Arka kiran, Arka mridula,
Karonda	Maroon
Arecanut	Mohit nagar
Acid lime	Rasraj
Coconut	Kera sankara, Chandra laksha
Litchi	Swarna roopa
Pomegranate	Ruby, Ganesh
Sapota	Cricket ball, Murrabba
Brinjal	Pusa kranti, Pusa uttam, Pusa upkar, Pusa ankur, Pusa hybrid-9, Arka anand, Arka keshav, Arka neel kanth, Arka nidhi, Arka sirish, Punjab jamuni gola, Punjab sadabahar, Punjab neelam, Pusa purple round, Black beauty
Pumpkin	Arka chandan, Arka suryamukhi
Okra	Arka abhay, Punjab padmani, Arka anamika, Parbhani kranti
Bottle gourd	Pusa samridhi, Arka bahar, PBOG-1, Azad nutan
Bitter gourd	Arka arupama, Arka harit
Tomato	Pusa gaurav, Pusa uphar, Pusa hybrid-1, Hemant, Pusa sadabahar, Sel-120, Pusa rohini (DT-39), Pusa ujwal, DTH-41, Arka samrat, Arka saurabh, Arka vardan, Arka vishal, Arka vikas, Pusa ruby, Pusa divya, Sun-496, Arka abhijeet, Arka abha, Pusa-120
Potato	Gauri sankar, Ree Bhadra, Kufri Swarna
Tuberose	Rajat Rekha
Cassava	Sree jaya, sree vijya
Cauliflower	Pusa deepali, Pusa sharad, Pusa subhra, Pusa snow ball, Pusa synthetic, Pusa meghna, CH-541, Janavon, Arka kranti,
Cabbage	Golden acre, Pusa drum head, Pusa mukta, All season, Red drum head, Express mail
Capsicum	Arka mohini
Coriander	Arka isha
radish	Arka nishant
Chilli	Pusa sadabahar, Pusa jwala (NP-46-A × Pusa red), Arka harita
Onion	Pusa red, Pusa madhvi, Pusa white flat, Pusa white round, Sel-402, Arka akshay, Arka bheem, Arka kalyan, Arka kirthiman, Arka lalima, Arka pitamber, Arka sona, Arka swadista, Arka ujjwal, Arka vishwas, Arka pragati
Ash gourd	Pusa ujwal, LDAG-4
Sponge gourd	Pusa supriya, Pusa sneha, DSG-5
Long melon	Arka sheetal (sel-3)
Musk melon	Arka jeet
Ridge gourd	Arka sujath, Arka Sumeet
Ginger	Surubhi, Suprabha
Turnip	Pusa chandrima, Pusa kanchan, Golden ball

Crop Name	Varieties
Sugarbeet	Crimson glove, Pusa kanchan
Carrot	Pusa keshar, Pusa meghali, Selection-233
Palak	Pusa jyoti, Arka anupama
Amaranthus	Arka suguna, Arka varna, Arka samraksha
Cucumber	Pusa sanyog, Pusa uday, DC-1
Snap melon	Pusa shandar
Water melon	MHW-6
Pumpkin	Pusa vishwash, Pusa vikas
Bougainvillea	Pusa tara, B.P. pal, Soonnet, Vishakha, Stanza, Summer time, Spring festival, Archana, Sweta
Rose	B.P. Pal, Jawahar, Pusa gaurav, Mrunalini, Chitra, Pusa priya,Pusa anhar, Pusa muskan, Pusa urmil, Pusa komal, Pusa shatabdi, Pusa abhishek, Pusa ajay, Pusa ranjana, Pusa Pitamber, Mother teresa, Bhim, Raktima, Mrinalini, Raktagandha, Pusa bahadur, Pusa mohit, Pusa arjun, Arka parimala, Priyadarshani
Chrysanthemum	Rakhee, Vasanti, Jaya, Ravi kiran, Rajat rekha, Apana, Indira, Red gold, Arka pink star, Nilima, Indira, Pusa anmol, Pusa centuary, Diana
Gladiolus	Shagun, Happy end, Poonam, Oscar, Suchitra, Shrun garika, Urvashi,Urmi, Jyotsana, Gulaal, Shabnam, Arka amar, Arka gold, Arka naveen, Kum kum
Marigold	Bolero, Golden jam, Lemon jam, Pumila, Pusa narangi
Opium poppy	Sweta, Kirtiman, Chetak, Trishna, Syama
Lemon grass	Kaveri, Praman, Sugandhi, Pragati
Palmariosa oil grass	RH-49, IW-31245
Carnation	Arka flame, Pusa hybrid-1
China aster	Kamini
Gerbera	Arka krishika, Panama
Ashwagandha	Poshita
Isabgol	Mandsaur-1, Pamarosa

Origin of Horticultural Crops

Origin	Crops Name
India	Kagzi lime, Bael, Jack fruit, Phalsa, Cucumber, Cluster bean, Pointed gourd, Snake gourd,
China	Litchi, Sweet orange, Mandrin, Persimon
Peru	Guava
Brazil	Pineapple, Cashewnut
South east Asia	Banana, Coconut, Lemon
Afghanistan	Almond, Carrot
Indo- Burma	Mango, Brinjal
Indi-China	Aonla, Palak,
Central Asia	Onion, Garlic, Pea

Origin	Crops Name
Europe	Raddish,
Tropical Africa	Musk Melon, Snap Melon, Long Melon, Water Melon, Round Melon
Maxico	Chilli, Pumpkin, Summer squash,
Africa	Okra,
South America	Potato, Tomato,
Mediterranean	Cabbage, Beet root,

Fruit Types of different Horticulture Crops

Fruit Type	Fruit/Vegetable
Drupe	Mango, Coconut, Almond, Ber, Phalsa, Date palm, Peach, Jamun, Olive
Nut	Cashew nut, Litchi, Water chestnut, Walnut, Peacanut,
Hesperidium	Lime, Sweet orange,
Berry	Brinjal, Tomato, Date palm, Banana, Papaya, Betel nut, Jamun, Grape, Guava, Karonda, Aonla, Avacado, Carambola,Sapota
Sorosis	Pineapple, Mulberry, Jackfruit
Syconus	Fig
Pepo	All cucurbits
Etaerio of Achenes	Strawberry
Lomentum	Tamarindus
Balausta	Pomegranate
Pome	Apple, Pear
Amphisarca	Bael
Capsule	Okra

Methods of propagation in different Horticulture Crops

Propagation Methods by	Crop Name
Seeds	Coconut, Papaya, Coffee, Phalsa
Seeds/Air layering	Citrus
Hard wood cutting	Grape, Fig,
Hard wood cutting/Air layering	Pomogranate
Sword suckers	Banana
Air layering	Litchi
Stooling	Guava
Inarching	Sapota
Runners	Strawberry
Suckers	Pineapple
Root cutting, Vine cutting	Sweet potato

Propagation Methods by	Crop Name
T budding/Patch	Aonla
T budding	Custard apple, Olive
Shield budding	Apple,Rose
Patch budding	Bael
T budding and Ring	Ber
Veneer grafting	Mango

Photoperiod of different Fruit and Vegetable Crops

Short day plant	Cluster bean, Sweet potato
Long day plant	Onion, Radish, Carrot, Potato, Cabbage, Palak, Turnip
Day neutral plant	Tomato, Chilli, Brinjal, Cucurbits, Cow pea, Okra

Edible Part of different Fruit and Vegetable Crops

Edible Part	Crop Name
Mesocarp	Papaya, Mango, sapota,
Pericarp	Ber, Date palm, Avacado, Custard apple, Tamarind
Endocarp	Coconut
Mesocarp and Endocarp	Banana, Aonla, Apricot, Watermelon, Muskmelon
Mesocarp and Epicarp	Jamun, Karonda, Plum, Peach
Thalamus and Pericarp	guava
Fleshy thalamus	Apple, Strawberry, Fig, Pear
Fleshy aril	Litchi
Juicy placental hairs	Citrus
Perianth/Bracts	Pineapple, Jackfruit
Pericarp and Placenta	Grape, Tomato, Brinjal, Chilli
Cotyledon	Cashew, Almond, Pecanut
Succulent placenta	Bael
Seed	Water chestnut,
Fruit	All cucurbitaceae family, Okra, Tomato, Brinjal, Chilli,
Bulb	Onion
Leaves	Fenugreek, Coriander, Amaranthus, Spinach,
Tuber	Potato, Cassava
Cloves	Garlic
Head	Cabbage
Curd	Cauliflower
Root	Carrot, Turnip
Pod	Cow pea
Pod and Seed	French bean, Cluster bean,
Adventitious root	Sweet potato

Famous Name of different Crops

Famous Name	Common Name
Coarsest of all food grains	kodo
King of cereals	Wheat
Queen of cereals	Maize
King of pulses	Chick pea
Queen of pulses	Pea
King of oilseeds	Groundnut
Queen of oilseeds	Sesame
King of fodder crops	Berseem
Queen of fodder crops	Lucerne
National fruits of India	Mango
King of fruits, Bathroom fruit	Mango
Apple of tropics	Guava
Queen of fruits	Pineapple/Mangosteen
King of temperate fruits	Apple
King of Arid and semi arid fruits	Ber
Indian black berry	Jamun
Adam fig, Kalpataru, Apple of paradise	Banana
Oldest cultivated tropical fruits	Banana
Tree of heaven, Kalpavriksha	Coconut
King of nuts	Walnut
Queen of nuts	Pecanut
Butter fruit	Avacado
Star apple	Phalsa
Queen of beverage crop	Tea
Star fruit	Carambola
Fancy fruit	Mandarin
King of vegetables	Potato
Queen of vegetables, lady`s finger	Okra
Egg plant	Brinjal
Vilayati baigan	Tomato
Vilayati kaddu	Summer squash
Vegetable of immense value	Pumpkin
King of flowers	Rose
Queen of flowers	Gladiolus
King of spices	Black pepper
Queen of spices	Cardamom
White gold of America	Cotton
Yellow jewel of America	Soybean
Backbone of America	Maize

Famous Name	Common Name
Sugar bowl	Cuba
Bio energy plant	Jatropha
Wonder crop	Soybean
Wonder tree	Neem
Miracle fruit	Kiwi
King of forest	Teak

Nutrient Content of Common Fertilizer

Fertilizer Name	N per cent	P2O5 per cent	K2O per cent	S per cent	Ca per cent	Mn per cent	Zn per cent
FYM (Farm Yard Manure)	0.5	0.2	0.5				
Groundnut cake	7.2	1.5	1.3				
Sunflower cake	7.8						
Blood meal	13-20						
Fish Meal	4-10	3-9					
Cotton cake	6.5						
Safflower cake	7.8						
Row Bone Meal		20-25					
Farm compost	0.5	0.15	0.5				
Poultry manure	3.0	2.6	1.4				
Nitrogenous fertilizer							
Nitrate form							
Sodium nitrate	16.0	-	-	-	-	-	-
Calcium nitrate	15.5	-	-	-	-	-	-
Ammonical form							
Ammonium chloride	26						
Ammonium sulphate	20-21	-	-	24			
Anhydrous Ammonia	82						
Ammonium phosphate	16	20					
Amide form							
Urea	46						
Calcium cyanamide	20.6						
Ammonical nitrate form							
Calcium Ammonium Nitrate (CAN)	26						
Ammonium Sulphate Nitrate	26			15			
Ammonium Nitrate	33-35						
Phosphatic fertilizer							

Fertilizer Name	N per cent	P2O5 per cent	K2O per cent	S per cent	Ca per cent	Mn per cent	Zn per cent
Water soluble							
SSP (Single Super Phosphate)	-	16	-	12	18-21		
DSP (Double Super Phosphate)		32					
TSP (Triple Super Phosphate)		48					
Di ammonia Phosphate (DAP)	18	46					
Citric acid soluble							
Dicalcium Phosphate (DCP)	-	33-40					
Basic Slag		14-18					
Bone meal		23-30					
Insoluble							
Rock Phosphate	-	20-30					
Rock bone meal		20-25					
Steamed bone meal		22					
Potassic fertilizer							
MOP (Murate of Potash)/KCl	-	-	60	-	-		-
Potassium Sulphate or Sulphate of Potash	-	-	48-52	17-18			
Potassium Nitrate	13	-	44				
					41-68		

Conversion Factor

$NO_3 = N \times 4.54$

$N \text{ (per cent)} = NO_3 \times 0.22$

$P = P_2O_5 \times 0.44$

$P_2O_5 = P \times 2.29$

$K = K_2O \times 0.83$

$K_2O = K \times 1.20$

$Ca = CaO \times 0.71$

$CaO = Ca \times 1.40$

$Mg = MgO \times 0.61$

$MgO = Mg \times 1.63$

Organic Matter (OM) = Organic Carbon (OC) $\times 1.72$

1000 ppm solution means 0.1 per cent or 1 g/litre water for spray

To convert Celsius into Fahrenheit :-

Multiply by 9, than devide by 5 and than add 32.

Ex. – $10^{\circ}C \times 9 = 90 \div 5 = 18 + 32 = 50^{\circ}F$

To convert Fahrenheit into Celsius:-

Subtract 32, than multiply by 5, and than divide by 9.

Ex.- $86°F-32=54×5=270÷9=30°C$

Nutrient Deficiency and Symptom in Plants

Symptom and Crop Name	Deficiency of Nutrient
Buttoning in cauliflower	N
Sickle leaf disease	P
Scorching and burning of leaves	K
Blossom end rot (BER) in tomoto	Ca
Tip hooking in cauliflower	Ca
Sand drawn of tobacco	Mg
Tea yellow disease	S
White eye of paddy	Fe
Intervienal chlorosis/leaf bleaching in sugarcane	Fe
Speckled yellow in sugarbeet	Mn
Pahala blight of sugarcane	Mn
Marsh spot of pea	Mn
Fruit cracking and rough bark in apple	Cu
Die back of shoot and Little leaf of citrus	Cu
Reclaimation disease in cereals	Cu
Whiptail of cauliflower	Mo
Khaira disease of paddy	Zn
White bud of maize	Zn
Little leaf of cotton, Mango, Brinjal	Zn
Bronzing in guava	Zn
Fruit cracking in tomato	Bo
Browning of cauliflower	Bo
Top sickness of tobacco	Bo
Snake heads in walnut	Bo
Internal necrosis in mango and Aonla	Bo
Hen and chicken disorder in grape	Bo
Heat rot in sugarbeet	Bo
Black heart of potato	O_2
Tip burn of paddy	O_2 deficiency and excess of Zn

Antagonistic Effect

Excess of Nutrient	Causes Deficiency
Ca	P
Ca, Mg	K
K, NH_4	Mg
Fe, SO_4	Mo
P	Zn
N, P, K	Cu
NO3	Fe
Zn, Al	Cu
N, K, Ca	B

Nutrient, available Form and First Appeared Plant Part

Nutrient	Available form	First Appeared Plant Part
N	NO^{-3}, NH_4^+	Older leaves
P	H_3PO_4, HPO^-_4, HPO_4^{-2} (acidic soil)	Older leaves
	PO_4^{-3} (alkali soil)	
K	K^+	Older leaves
Mg	Mg^{+2}	New leaves
Mn	Mn^{+3}, Mn^{+2}	Older leaves
Mo	MO^{+3}, MoO^-_4	New leaves
Zn	Zn^+	Small yellow in new leaves
Ca	Ca^{+2}	Young leaves/Terminal buds
S	SO_3^{-3}, SO_4^{-2}	First in new leaves
Fe	Fe^{+2}, Fe^{+3}	New leaves
Cu	Cu^+, Cu^{+2}	New leaves
B	Bo_3^{-3}, BO_4, $H_2Bo_3^-$, HBO_3^{-2}, $HB_4O^-_3$	New leaves
Cl	Cl^-	
CO	CO^{+2}	
C	CO_3^{-2}, HCO_3^{-2}	
O, H	H^+, OH^-	
H_2O	H^+, OH^-	

Indicator Crops

Deficient Nutrients	Indicator Crops
Nitrogen (N)	Maize, Mustard, Apple, Citrus, Non-leguminous small grains
Phosphorus (P)	Maize, Tomato, Barley, Luttuce, Sugarbeet
Potassium (K)	Tomato, Cotton, Cucurbits, Clover, Lucern, Potato, Tobacco, Bean, Sugarcane

Deficient Nutrients	Indicator Crops
Calcium (Ca)	Lucerne, Berseem
Magnesium (Mg)	Potato, Cauliflower
Iron (Fe)	Sorghum, Citrus, Barley, Peach, Spinach, Sugarbeet, Tomato, Banana
Sulphur (S)	Clover, Lucern, Berseem, Soybean, Tobacco
Copper (Cu)	Maize, Wheat, Luttuce, Oat, Tobacco, Tomato, Apple, Bean, Apricot, Ber, Onion, Litchi, Citrus
Boron (Bo)	Cauliflower, Turnip, Lucern, Apple, Peach
Manganese (Mn)	Maize, Oat, Pea, Wheat, Bean, Cherry, Apple, Apricot, Radish
Chlorine (Cl)	Lettuce, Maize, Tomato
Molybdenum (Mo)	Lobia, Lucerne, Sugarbeet, Cauliflower, Tomato
Zinc (Zn)	Paddy, Wheat, Maize, Cotton

Cross Inoculation Group of Rhizobium

Rhizobium Species	Cross Inoculation
Rhizobium trifoli	Clover group
Rhizobium meliloti	Alfa group
Rhizobium phaseolin	Bean group
Rhizobium lupini	Lupine group
Rhizobium leguminosarum	Pea group
Rhizobium japonicum	Soybean group

Toxic Substances found in different Crops

Toxic Substances	Crops
Hydrocyanic acid (HCN) or Prussic acid	Sorghum
Oxalic acid	Napier grass, Pearl millet
Mimosine	Leucarena leucocephala
Saponins	Clover, Alfalfa, Strawberry
Castrogens	White clover
Coumarins	Sweet clover
Erusic acid	Mustard, Rape seed
Trypsin	Pigeon pea, Cowpea
Cyanogenic glucosides	Sorghum, Sudan grass
Polyphenolics	Safflower
Lathyrogen or Neurotoxin	Lathyrus

Different Agroforestry system

Agrisilviculture	Agriculture crops + Forest trees
Silvipastoral	Forest trees + Grasses
Agrisilvipastoral	Agriculture crops+ Forest trees+ Grasses
Agrihorti system	Agriculture crops+Fruit crops
Hortisilviculture	Fruit crops+ Forest trees
Agrohortisilviculture	Agriculture crops+ Fruit crops+ Forest trees
Aqualsilviculture	Fish+Forest trees
Agrisilviaquaculture	Agriculture crops+Forest trees+Fish

Instruments and their Measurement Use

Instrument	*Measures*
Anemometer	Wind Velocity
Auxanometer	Growth of plant
Aneroid barometer	Atmospheric pressure
Altimeter	Height
Almometer	Evaporation rate
Aerometer	Weight and density of air and gases
Actinometer	Solar radiation
Barometer	Atmospheric pressure
Barograph	Atmospheric pressure graph
Beaufort scale	Wind pressure is measured by
Brix hydrometer	Brix of sugar cane juice
Butyrometer	Fat content in milk or milk product
Crescograph	Growth of plant in graph
Cambel stokes	Sunshine duration
Cryometer	Temperature below 0°c
Drosometer	Dew
Fathometer	Depth of sea
Hydrometer	Relative density of liquid
Psychrometer/Hygrometer	Relative humidity
Irrometer	Water stress, soil moisture tension
Infiltrometer	Infiltration
Infra red Thermometer	Crop canopy temperature
Lactometer	Fat percentage in milk/specific gravity of milk
Lysimeter	Evapotranspiration/percolation and leaching
Manometer	Root pressure
Porometer	Stomatal actvity
Photometer	Transpiration rate
Photometer	Luminous intensity of light

Instrument	Measures
Pyrheliometer	Amount of direct solar radiation
Pycnometer	Specific gravity of soil
Pyranometer	Total incoming solar radiation/very high temperature
Quantum sensor (Lux is oldest unit)	PAR (photo synthetic active radiation)
Psychrometer	Leaf water potential/relative humidity
Piezometer	Depth of water level
Planimeter	Area of an irregular figure
Tensiometer	Soil moisture tension
Wind vane	Wind direction

Equal Line Terminologies

Line of equal number of neutron	Isobar
Line of equal number of electron	Isotope
Line of equal moisture	Isotere
Line of equal wind speed	Isotach
Line of equal atmospheric pressure	Isoline
Line of equal amount of rainfall	Isohyets
Line of equal water table	Isobath
Line of equal temperature	Isotherm
Line of equal depth of rainfall	Isopluviel
Line of equal rainfall	Isohytal
Line of equal sunshine hours	Isochrones

Chromosome Number of Reproductive Parts

Pollen grain, Pollen tube, Megaspore, Microspore	= n
Seed, Embryo, Pollen mother cell, Megaspore mother cell	= 2n
Endosperm	= 3n

Isolation Distance

Crop Name	Foundation Seed (meter)	Certified Seed (meter)
Paddy, Barley, Wheat, Oat, Groundnut, Soybean	3	3
Hybrid maize, Berseem, Lucern, Capsicum, Okra, Sunflower	400	200
Gram, Urad, Moong, Lentil, Pea	10	5
Arhar, Brinjal	200	100
Cotton, Jute, Tomato, Bean, Lobia	50	25
Mustard, Til	100	50
Castor	300	150
Potato	20	5

Crop Name	Foundation Seed (meter)	Certified Seed (meter)
Onion, Carrot, Cucurbits	1000	500
Spinach, Cauliflower, Cabbage, Radish	1600	1000
Cucurbits (hybrid)	1000	1500

Types of Larva

Order	Larva
Hemiptera (bugs, hoppers, white fly, aphids, jassids)	Nymph
Lepidoptera (borers, moths, bollworms)	Catterpiller
Coleoptera and Hymenoptera (mustard saw fly)	Grub
Diptera (all true flies)	Maggot

Types of Mouth Parts of Insects

Mouth Part Type	Insects Name
Biting and chewing type	Grass hopper, Locust, Beetles, Lepidoptera larva
Piercing and sucking type	Leaf hopper, Mosquito, Aphids, Bugs
Sponging type	House fly
Siphoning type	Butter fly, Moths
Rasping and lapping type	Honey bees

Modification of Insect Antennae

Type of Modification	Insect Name
Aristate	Housefly
Bipectinate	Silkworm
Clavate	Butter fly
Filiform	Grasshopper, Ground beetle
Geniculate	Honey bee
Moniliform	Termite
Plumose	Male mosquitoes
Pectinate	Moths, Sugarcane root borer

Modification of Insect Legs

Type of Modification	Insect Name
Climbing type	House fly
Clasping type	Pyrilla
Digging type	Male cricket

Type of Modification	Insect Name
Jumping type	Grass hopper, Cricket
Pollen collecting type	Honey bee
Running type	Ants
Raptorial or Grasping type	Mantis
Sound producing type	Cricket, Hind leg of male grass hopper
Swimming type	Dytiscus and joint water bug
Walking type	Cockroach and Bugs

List of Major Crop Diseases

Crop Name	Disease Name	Causal Organism
Paddy	Blast (Air borne/Rich man's disease)	*Pyricularia oryzae*
	Brown spot (Externally seed borne)	*Helminthosporium oryzae*
	Stem rot	*Sclerotium oryzae*
	Sheath blight (soil borne)	*Rhizoctonia solani*
	Sheat rot	*Sarocladium oryzae*
	Bacterial leaf spot (Poor man's disease)	*Xanthomonas oryzae*
	Flase smut	*Claviceps oryzae*
Wheat	Stem rust/black rust	*Puccinia graminis tritici*
	Leaf/orange/brown rust	*Puccinia recondita*
	Strip/yellow rust	*Puccinia striformis*
	Loose smut(Internally seed borne)	*Ustilago tritici*
	Flag smut	*Urocystis tritici*
	Karnal bunt (Air, soil and seed borne)	*Neovassia indica*
	Leaf blight	*Alternaria triticina*
	Alternaria blight	*Alternaria triticina*
	Ear cockle (by nematode)	*Aguena tritici*
Sorghum/ Jowar	Anthracnose	*Colletotrichum graminicolum*
	Grain smut	*Sphacelotheca sorghi*
	Loose smut	*Sphacelotheca cruenta*
	Head smut	*Sphacelotheca reiliana*
	Long smut	*Tolyposporium ehrenbergii*
	Ergot sugary disease	*Sphacelia sorghi*
Pearl millet/ Bajra	Downy mildew	*Sclerospora graminicola*
	Rust	*Puccinia penniseti*
	Smut	*Tolyposporium penicilliariae*
	Ergot	*Claviceps fusiformis*
Pea	Wilt (soil borne)	*Fusarium oxisporium*
	Collar rot	*Aspergillus niger*
	Powdery mildew	*Erisyphe poligoni*
	Sclerotinia blight	*Sclerotinia sclerotiorum*

Crop Name	Disease Name	Causal Organism
Arhar	Wilt (soil borne)	*Fusarium oxisporium f. spp. udum*
Cotton	Wilt	*Fusarium moniliform*
Sugarcane	Red rot	*Colletotrichum falcatum*
	Smut	*Ustilago scitaminea*
	Wilt	*Cephalosporium sacchari*
	Rust	*Puccinia erianthi*
	Downy mildew	*Peronosclerospora sacchari*
	Eye spot	*Helminthosporium sacchari*
	Ring spot	*Leptosphaeria sacchari*
	Gummosis	*Xanthomonas campesris vasculorum*
	Red strip	*Pseudomonas rubrilineans*
Groundnut	Early leaf spot/Tikka	*Cercospora arachidicola*
	Late leaf spot/Tikka	*Cercospora personata*
	Collar rot	*Aspergillus niger*
	Rust	*Puccinia arachidis*
	Anthracnose	*Colletotrichum dematium*
Barley	Covered smut	*Ustilago hardei*
	Loose smut	*Ustilago nuda*
	Powdery mildew	*Erysiphe graminis* var. *hordei*
	Anthracnose	*Colletrotrichum indicum*
Mustard	Alternaria blight	*Alternaria brasicae*
	Downey mildew	*Pernospora brassicae*
	White rust blister	*Albugo candida*
Sunflower	Rust	*Puccinia helianthi*
	Downey mildew	*Plasmopara halatedil*
Maize	Brown spot	*Physoderma maydis*
	Smut	*Ustilago maydis*
	Leaf blight	*Helminthosporium maydis*
	Rust	*Puccinia sorghi*
	Downy mildew	*Peronosclerospora sorghi*
	Seed rot or seedling blight	*Fusarium monilliformis*
	Bacterial stalk rot	*Erwinia crotovorta* var. *zea*
	Charcoal rot	*Macrophomina phaseoil*
Tobacco	Damping off	*Pythium aphanidermaum*
Soybean	Anthracnose (Pod blight)	*Colletotrichum truncatum*
Cucurbits	Powdery mildew	*Erysiophe cichoracearum*
	Downey mildew	*Pseudopernospora cubensis*
Potato	Late blight	*Phytophthora infestans*
	Early blight (soil borne)	*Alternaria solani*
	Black scurf (Tuber borne)	*Rhizoctonia solani*
	Weart disease	*Synchutium endobioticum*

Crop Name	Disease Name	Causal Organism
Okra	Powdery mildew	*Erysiophe cichoracearum*
Apple	Scab	*Venturia inaequalis*
	Black canker	*Sphaeropsis malorum*
	Fire blight	*Erwinia amylovora*
Guava	Anthracnose	*Collectotrichum psidii*
Papaya	Stem and foot rot	*Pythium aphanidermatum*
Banana	Panama wilt	*Fusarium oxysporium* var. *cubense*
Citrus	Canker	*Xanthomonas compestris* pv *citri*
	Gummosis (Broom rot)	*Phytophthora palmivora*

Anti-transpirents

☆ **Stomatal closure type** – 2,4-D, Atrazine, Potassium metabisulphite, PMA.

☆ **Growth retardant type** – Cycocel (CCC), Phosphor.

☆ **Reflectant type** – Kaoline (5 per cent), Calcium bicarbonate, Lime water, China clay.

☆ **Film farming type** – Paclobutrazole, Hexadeconol, Cetyl alcohol, Waxol, Folicot, Silicon.

Functions of Plant Growth Regulator (PGR)

☆ **Auxin** – Cell division and root formation, Enhances fruit set and fruit ripening, can be used as herbicide.

☆ **Gibberellin** – Cell division, breaking dormancy and cell elongation, increasing size of fruit and flower, leaf and flowering induce.

☆ **Cytokinin** – stimulate cell division, delay senescence, breaking dormancy of seed and development of embryo in seed.

☆ **Abscisic acid** – Induce dormancy and maintain cell turgidity, facilitate stomata closure.

☆ **Ethylene** – Fruit ripening, iso-diametric growth of stems and root.

Revolution in Agriculture

Green Revolution	Wheat and Paddy Production
White revolution	Milk/Dairy Production
Yellow revolution	Oil Seeds Production
Blue revolution	Fish Production
Round revolution	Potato Production
Silver revolution	Egg/Poultry Production
Brown revolution	Leather/Cocoa Production/Non-Conventional Energy sources
Gray revolution	Fertilizer Production

Green Revolution	Wheat and Paddy Production
Red revolution	Tomato/Meat Production
Pink revolution	Prawn Production/Onion/Pharmaceutical
Golden revolution	Horticulture development/Honey production
Black revolution	Bio Fuel Production
Rainbow revolution	Agriculture Production
Evergreen revolution	Reduction in wastage of food grains, fruits and Vegetables/over all development of agriculture
Parbhani revolution	Okra Production
Golden fibre revolution	Jute production
Silver fibre revolution	Cotton production

Grain Straw Ratio

Crop Species	Harvested Component	Harvest Index	Grain/Straw Ratio
Wheat, Barley	Grain	0.55	1.22
Rice	Grain	0.50	1.11
Maize	Grain	0.52	1.16
Chickpea	Pods	0.30	0.43
Soybean	Pods	0.40	0.67
Sunflower	Seeds	0.50	1.11
Bean (Phaseolus)	Pods	0.25	0.56
Peanut	Pods	0.50	1.11
Cotton	Bolls	0.33	0.73
Sugarbeat	Root	0.50	1.11
Potato	Tuber	0.82	1.82
Sweet Potato	Tuber	0.65	1.44

Vitamin Sources, Deficiency Causes and Chemical Name

S. No.	Vitamin/ Nutrient Name	Also known	Deficiency cause	Highest Vitamin source by fruit	Highest Vitamin source vegetable
1.	Vitamin A	Retinal	Night blindness	Mango	Bathua
2.	Vitamin B_1	Thiamin	Beriberi	Cashew nut	Chilli
3	Vitamin B_2	Riboflavin	Skin cracking	Bael	Fenugreek
4.	Vitamin B_3	Pantathenic acid	Whiteness of hair		
5.	Vitamin B_5	Niacin	Pellagra		
6.	Vitamin B_6	Pyridoxine	Degration of nerves		
7.	Vitamin B_7	Biotin	Paralysis		

S. No.	Vitamin/ Nutrient Name	Also known	Deficiency cause	Highest Vitamin source by fruit	Highest Vitamin source vegetable
8.	Vitamin B_{12}	Cynocobaltamin	Pernicious anaemia		
9.	Vitamin C	Ascorbic acid	Scurvy	Barbados cherry	Drumstick leaves
10.	Vitamin D	Calciferol	Rickets		
11.	Vitamin E	Tocopherol	Sterility		
12.	Vitamin K	Phylloquinone	Non coagulation of blood		
13.	Carbohydrate			Raisins	Tapica
14.	Protein			Cashew nut	Lima bean
15.	Fat			Walnut	Potato
16.	Fibre			Fig	Potato
17.	Calcium			Litchi	Agathi
18.	Phosphorus			Almond	Amaranthus
19.	Iron	•	•	Karonda	Amaranthus

Biological Control

Bio-agent (Predators)	Family	Target
Lady beetle	Coccinillidae	Aphid, Spider mite
Groundnut beetle	Caradidae	Insect
Rove beetle	Staphylinidae	Small insect
Green lacewing	Chrysopidae	Caterpillar, Beetle
Syrphid fly	Syrphidae	Aphid
Predatory bugs	Hemiptera	Insect, Mites
Stink bugs	Pentatomidae	Potato beetle larvae
Damsel bugs	Nadidae	Aphids
Assassin bugs	Reduvidae	Caterpillar, Beetles
Minute pirate bugs	Anthocoridae	Thrips, Spider mites
Hunting wasps	Sphecidae	Caterpillar, Beetle
Spider	Araneida	Living insect, Small anthropods
Bio agent (Parasite)	Family	Target
Tachinid flies	Tachinidae	Caterpillar, Beetle, Bugs
Braconid wasps	Braconidae	Aphid, Small insect
Chalcid wasps	Chalcidoidae	Aphid

Important Disease of Animals

Disease Name	Caused by
FMD, Rinderpest, Cow pox, Ranikhet, Marek`s,	Virus
Liver fluke	Ascariasis
Ranikhet	Virus

Disease Name	Caused by
Marek`s	Virus
Anthrax, Foot rot, Fowl cholera	Bacterial
Coccidiosis	Protozoal

General Comparison of Animals

Animal name	Group is known as	Meat	Adult male	Adult female	Gestation Period (Days)	Pulse Rate/min	Body Temp. (°F)
Cattle	Herd	Calf–Veal Cattle–Beef	Bull	Cow	283-285	50-60	101.5
Buffalo	Herd	Buffen	Buffalo bull	Buffalo	300-310	40-50	98.3-101
Poultry	Flock	Chicken	Cock	Hen	21	120-160	107
Goat	Flock	Chevon	Buck	Doe	148-150	70-90	103.8
Sheep	Flock	Mutton	Ram	Ewe	149-150	70-90	102
Pig	Herd	Pork	Boar	Sow	114	70-80	102.9

Different Magazine and Publication

Magazine Name	Published by/Published from
Kurukshetra	Ministry of Rural Development
Yojna	Ministry of Information and Broadcasting
Mausam	Indian Meteorological Department
Science Reporter	CSIR
Agriculture Extension Review	Ministry of Agriculture
Unnat Krishi	Ministry of Agriculture
Agriculture situation in India	Ministry of Agriculture
Agriculture statistics at a glance	Ministry of Agriculture
Intensive Agriculture	Ministry of Agriculture
Krishi Vistav Samiksha	Ministry of Agriculture
Indian Farmer Digest	GBPAU
Fertilizer News	Fertilizer Association of India
Annuals of Agriculture Research	Indian Society of Agriculture Science
Rural India	Pune
Communicator	Indian Institute of Mass Communication
Prasarika	Rajasthan Society of Extension Education
Extension Education in community development	Ministry of Food and Agriculture
Encyclopaedia of social work in India	Planning Commission
Extension digest	MANAGE
Social action	Institute of Social Science, New Delhi

Different National Boards

National Boards Name	Establishment year	Situated at
Biodiversity board	2008	Chennai
Central silk board	1949	Banguluru
Coffee board	1942	Banguluru
Coconut development board	1981	Kochi
Fish and fish product board	2006	Hyderabad
Rubber board	1947	Kottayam, Kerala
Spice board	1986	Kochi
Tea board	1953	Kolkata
Tobacco board	1976	Guntur,A.P.

Different Act/Order/Policy

Act/Order	Passed Year
Agriculture Refinance and Development Corporation Act	1963
Agriculture Produce (Grading and Marketing) Act	1937
Biological Diversity Act	2002
Bonded Labour System Act	1976
Bureau of Indian Standard Act	1986
Copyright Act	1957
Copyright Act (Amendment)	1999
Corporation Act	1962
Central Pollution Control Board	1974
Cattle Trespass Act	1871
Consumer Protection Act	1986
Drug and Cosmetics Act	1940
Destructive Insects and Pest Act	1914
Energy Conservation Act	2001
Employer`s Provident Fund and Family Pension Act	1972
Essential Commodities Act	1955
Equal Remuneration Act	1976
Environment Protection Act	1986
Fertilizer Control Order	1985
Forest Survey of India Act	1981
Forest Conservation Act	1980
Fertilizer Movement Control Order	1969
Fruit Products Order	1953
Food Security Bill	2011
Insecticide Rule	1971

Act/Order	Passed Year
Indian Forest Act	1927
Insecticide Act	1968
Indian Fisheries Act	1897
Insecticide (Price, Stock, Display and Submission of Reports) Order	1986
Jawaharlal Nehru Solar Mission	2010
Land Improvement Act	1871
Locust Warming Organization	1939
Land Acquisition Act	1894
Livestock Importation Act	1898
Milk and Milk Product Order	1992
Meat Food Product Order	1973
Minimum Wages Act	1948
Micro Irrigation Scheme	2006
National Seed Policy	2002
National Biodiversity Act	2002
National Water Policy	2002/2012
New and Renewable Energy Policy	2005
National Policy on Biofuels	2009
Plant Quarantine Order	2003
Protection of Plant Varieties and Farmers Right Act	2001
Plant, Fruit and Seed Order	1989
Payment of Bonus Act	1965
Prevention of Food Adulteration Act	1955
Payment of Wages Act	1936
RBI Act	1934
Regional Rural Bank Act	1976
Royal Commission Report	1928
Standard of Weight and Measures Act	1976
Seed Act	1966
Seed Rule	1968
Seed Control Act	1983
Vegetable Oil Products Control Order	1977
Water (Prevention and Control of Pollution) Act	1974
Warehousing Corporation Act	1962
Wildlife Production Act	1972
Zoological Survey of India	1916

Different Theories

Theories	Economist
Theories of Profit	Walker
Theory of Rent	Ricardo
Theories of Consumption	Keynes
Morden Theory of Wages and Employment	Keynes
Theories of Time Preference	Fisher
Theories of Employment	Keynes
Morden Theories of Interest	Hicks – Hansen
Wage Fund Theory	J.S. Mill
Population Theory	Malthus
Theories of Multiplier	Keynes
Theories of Disguised Unemployment	Nurkeses
Theories of Unlimited Supplies of Labour	Lewis
Low Level Equilibrium Trap	Nelson
Theories of Undeveloped Countries	Keynes
The stages of economic growth and classical theory of economic development	W.W. Rostow
Theories of Das Kapital	Karl Marks
Theories of Economic Growth	John fei and G. Rains
Theories of Big Push	N. Rosenstein Rodan
Theories of Social Dualism	J.H. Boeke
Theories of Circular Causation	Myrdal
Requirement of Steady Growth	Harrod – Domar
The Model of Profit and Growth	Psinette
The Model of Capital Accumulation	John Robinson
The Model of Economic Growth	R.E. Meade
The Model of Growth	Kaldor
The Single Sector Model	Mahalanobis
Theories of Liquidity Preference	Keynes
Monetary Theory	R.G. Howtrey
Sunspot Theory	Stanley jevon
Under Consumption Theory	J.A. Hobbson
Quantity Theories of Money	Millon Friedman
Theories of Absolute Advantage	Adam Smith
Theories of Inflation	A.P. Herner
Modern Theories of International Trade	Ohlim

Krishi Karman Award (Govt. of India)

Category	State (2011-12)	State (2014-15)	2015-16
For Big state	Punjab and Uttar Pradesh	Madhya Pradesh	Tamil Nadu
For Medium state	Odisha and Assam	Odisha	Himachal Pradesh
For Small state	Tripura	Meghalaya	Tripura
For Paddy	Chhattisgarh	Haryana	—
For Wheat	Haryana	Rajasthan	Madhya Pradesh
For Pulse	Maharashtra and Rajasthan	Chhattisgarh	—
For Coarse cereals	Karnataka	Tamil Nadu	—

Leading State in Production and Area of Crops

Paddy production	WB
Paddy area	WB
Paddy productivity	Punjab
Wheat production	UP
Wheat area	UP
Wheat productivity	Haryana
Pulse production	MP
Pulse productivity	Haryana
Oil seed production	MP
Oil seed productivity	TN
Groundnut production	Gujarat
Groundnut productivity	TN
Mustard production	Raj
Cotton production	MH
Jute production	WB
Tea production	Assam
Rubber production	Keral
Potato production	UP
Onion production	MH
Sugarcane production	UP
Sugarcane productivity	TN
Maize production	Karnataka
Soybean production	MP
Soybean productivity	AP
Fruit crop leading in Area	Mango
Fruit crop leading in Production	Banana
Fruit crop leading in Productivity	Papaya
State leading in fruits crops area	MH
State leading in fruits crops production	AP

Vegetable crop leading in area	Potato
Vegetable crop leading in production	Potato
Vegetable crop leading in productivity	Tapioca
State leading in vegetable crops area	WB
State leading in vegetable crops production	WB

India Position in World (2016-17)

Content	India Rank	1st Rank Conutry
Total geographical area	7	-
Irrigated area	1	India
Human Population	2	China
Total cereals	3	China
Wheat, Paddy	2	China
Coarse grain	4	USA
Total pulse	1	India
Oil seed	2	-
Wheat	3	Europe Union
Vegetable	2	China
Fruit	2	China
Milk	1	India
Maize	4	USA
Groundnut	2	China
Sugarcane	2	Brazil
Cotton, Tobacco	3	China
Tea, Jute	7	India
Coffee	6	-
Milk production	1	India
Egg production	4	China
Plant biodiversity	4	-
Mustard	3	China

Important Project and Associated Persons

Project Name	Associated Person
Shrineketan	R. N. Tagore
Gurgon Project	F. L. Baryne
Seva Gram	M. K.Gandhi
Marthandam	Spencer Hatch
Etawah Pilot Project	Albert mayer
Nilikhedi Experiment	S. K. Dey

Project Name	Associated Person
Rural Reconstruction	Denial Hamilton
Majdoor Manjil	S. K. Dey
Firaka	T. Prakasam
Community Development	S. K. Dey
Indian Village scheme	S. N. Gupta
Young Farmer Association	Deshmukh
National demonstration Scheme	Kalwar and Subramanyam

Important Programme and Starting Years

Programme Name	Starting Year
Royal commission	1928
ICAR establishment	1929 July 16
RBI establishment	1935 April 1
AGMARK	1937
Grow more food compaign (In 1952 GMF enquiry committee was constituted)	1947
GATT set up (Replaced in 1995 by WTO)	1947
RBI Nationalised	1949 January 1
Planning Commission	1950
Community Development Programme (CDP) (55 CD block)	1952 October 2
National Development Council (NDC)	1952
National Extension Service (NES)	1953 October 2
All India Co-Ordinated Research Project On Maize	1957
Village Panchayat Act	1958
PERT	1958-59
Panchayati Raj System (B.R. Mehta 3-teir)	1959 October 2
Intensive Agriculture Area Programme (IADP)	1960 Kharif
1st Agriculture university on land grant pattern at Pantnagar	1960
Intensive Agriculture Area Programme (IAAP)	1964
FCI Act	1964
Food Cooperation of India (FCI)	1965 January
Agriculture Price Commission (APC)	1965
National Dairy Development Board (NDDB)	1965
National Demonstration Programme	1965-66
High Yielding Variety Programme (HYVP)	1966 Kharif
Multi Cropping Scheme	1966
Green revolution (during 3 rd five year plan)	1966-67
Television Start As Communication Media/Krishi Darshan	1967
14 Commercial Bank Nationalised	1969 July 19
Lead Bank Scheme Introduce	1969

Programme Name	Starting Year
HUDCO	1970
SFDA and MFDA	1970-71
Drought Prone Area Programme (DPAP)	1970-71
Operation Flood	1971
Department of Agriculture Research and Education (DARE)	1973 December
Department of Agriculture Research And Education (DARE) (Under Ministry of Agriculture)	1973
ASRB establishment	1973 November 1
Krishi Vigyan Kendra (KVK)	1974
Training and Visit System (T and V System) (1st KVK at Pondichery)	1974
Command Area Development Programme (CADP)	1974
Integrated Child Development Service (ICDS)	1974
Operational Research Project (ORP)	1974-75
20 Pointed Programme (Started by Indira Gandhi)	1975
National Seed Programme	1975
Regional Rural Bank (RRB)	1975 October 2
Antyodaya Programme	1977
Desert Development Programme (DDP)	1977-78
Food For Work Programme (but national food for work programme started on 2004)	1977-78
National Rural Employment Programme (NREP)	1980
Lab To Land Programme (Starting On Golden Jubilee of ICAR)	1979
TRYSEM	1979 August 15
National Agriculture Research Project (NARP)	1979-80
Food And Work Programme	1980
Integrated Rural Development Programme (IRDP) (1st time launched in 1978 in 2300 blocks and in 1980 in all blocks)	1980 October 2
District Rural Development Agency (DRDA)	1980
National Rural Development Guarantee Programme (NREGP)	1980
Development of Women and Children In Rural Areas (DWCRA)	1982
NABARD Act was passed	1979
NABARD	1982 July 12
Comprehensive Crop Insurance Scheme	1985
Transfer of Technology Programme	1985
Indira Awas Yojna (IAY) (Central :State=75:25)	1986-87
CAPART-1986	1986
SEBI	1988
Jawahar Rojgaar Yojna (JRY) (started by Indira Gandhi)	1989
Integrated Waste-Land Development Programme (IWDP)	1989-90
National Watershed Development For Rainfed Area (NWDPRA)	1990-91
National Watershed Development Programme (NWDP)	1991

Programme Name	Starting Year
Prime Minister Rojgaar Yojna (PMRY)	1993
IMF	1994
Mid-Day Meal Scheme	1995 August 15
Institutional Village Linkage Programme (IVLP)	1995
WTO	1995 January 1
National Gene Bank at New Delhi established	1996
Rural Employment Generation Programme (REGP)	1995
Swarn Jayanti Sahri Rojgaar Yojna	1997 December
National Agriculture Technology Project (NATP)	1998
ATIC	1998
Kisan Credit Card Scheme-	1998-99
Jawahar Gram Samridhi Yojna (JGSY) (Old Name JRY)	1999 April 1
District Rural Development Agency (DRDA)	1999 April 1
Swarn Jayanti Gram Swarojgar Yojna (SGSY) (Central :State 75:25)	1999 April 1
National Agriculture Insurance Scheme (NAIS)	1999 Rabi
Anotodaya Anna Yojna (AAY)	2000 December 25
Annapurna Scheme	2000
National Agriculture Policy	2000
Prime Minister Gram Sadak Yojan (PMGSY)	2000 December 25
Pradhan Mantri Gramodaya Yojna (PMGY)	2000-01
Sampurna Grameen Rojgaar Yojna (SGRY)	2001 December 25
Agriculture Technology Management Agency (ATMA) (2006 -07 in MP 15 and 2007-08 started in all MP all district)	2002
National Seed policy	2002
Community Radio Station	2002 December
Kisan Call Centre (KCC) (At present 25 KCC working in India)	2004 January 21
Integrated scheme on Oilseeds, Pulses, Oil palm and Maize (ISOPOM)	2004 April 1
Crop Agriculture Produce Loan Scheme (CAPL)	2005 April 1
National Horticulture Mission (NHM)	2005 May
5Th Economic census	2005
National Rural Health Mission (NRHM)	2005
Janani Suraksha Yojna	2005 April 12
NREGA (Minimum 100 days wok in a year)	2005 December 7
National Agriculture Innovation Programme (NAIP)	2006 July
National Rural Employment Scheme	2006
National Bamboo Mission	2006-07
Rashtriya Krishi Vikas Yojna (RKVY)	2007 May 29
18th Live Stoke Census (takes place after every 4 year)	2007 October 15
National Food Security Mission (NFSM)	2007 May 29

Programme Name	Starting Year
Rural Employment Guarantee Act.	2008 April 1
National Mission on Micro Irrigation (NMMI)	2010 June
15th human population census (After independence it's a 7th census and first census occurred in 1881)	2011

International Institutes and Research Centres

- ☆ International Crops Research Institute for Semi Arid Tropics (ICRISAT) – Hyderabad (India)
- ☆ International Centre for Agriculture Research in Dry Land (ICARDA) – Aleppo (Syria)
- ☆ International Centre for the Improvement of Maize And Wheat (CIMMYT) – Mexico
- ☆ International Rice Research Institute (IRRI) – Manila (Philippines)
- ☆ International Plant Genetic Resource Institute (IPGRI) – Rome (Italy)
- ☆ International Board for Plant Genetic Resource (IBPGR) - Rome (Italy)
- ☆ International Centre for Genetic Engineering and Biotechnology – Triesta (Italy)
- ☆ West Africa Rice Development Association (WARDA) – Monrovia (Liberia)
- ☆ International Centre for Tropical Agriculture (CIAT) – Coli (Columbia)
- ☆ International Centre for Research in Agro Forestry (ICRAF) – Nairobi (Kenya)
- ☆ International Livestock Research Institute (ILRI) – Nairobi (Kenya)
- ☆ International Centre for Living Aquatic Resource Management (ICLARM) – Manila (Philippines)
- ☆ International Potato Centre (CIP) – Lima (Peru)
- ☆ International Institute for Tropical Agriculture (IITA) – Ibadan (Nigeria)
- ☆ International Irrigation Management Institute (IIMI) – Colombo (Srilanka)
- ☆ Centre for International Forestry Research (CIFOR) – Bogor/Jakarta (Indonesia)
- ☆ International Food Policy Research Institute (IFPRI) – Washington (USA)
- ☆ International Network on Soil Fertility and Fertilizer Evaluation on Rice (INSFFER) – New Delhi
- ☆ International Institute of Rural Reconstruction (IIRS) - Philippines.
- ☆ International Service for National Agriculture Research (ISNAR) – Netherlands
- ☆ International Water Management Institute (IWMI) – Columbo (Sri Lanka)
- ☆ World Fish Centre (WFC) – Malaysia

- ☆ Asian Vegetable Research and Development centre (AVRDC) – Taiwan
- ☆ Consultative Group on International Agricultural Research (CGIAR) – Washington (USA)
- ☆ Food and Agricultural Organization (FAO) – Rome (Italy)
- ☆ International Bank for Reconstruction and Development (IBRD) - Washington (USA)
- ☆ International Benchmark Soils Network for Agrotechnology Transfer (IBSNAT) – Hawai
- ☆ International development Research Centre (IDRC) – Canada
- ☆ International Fertilizer Development Centre (IFDC) – Alabama (USA)
- ☆ International Institute of Tropical Agriculture (IITA) – Nigeria
- ☆ International Soils Reference and Information Centre (ISRIC) – Netherlands
- ☆ Land Resources Development Centre (LRDC) – UK
- ☆ Tropical Pesticides Research Institute (TPRI) – Tanzania
- ☆ World Meteorological Organization (WMO) – Geneva (Switzerland)

Important Days

☆ WTO Foundation Day	– 1 January
☆ World Hindi Day	– 10 January
☆ National Wetland Day	– 2 February
☆ World Cancer Day	– 3 February
☆ National Science Day	– 28 February
☆ Water Resource Day	– 11 march
☆ World Forestry Day	– 21 March
☆ World Water Day	– 22 March
☆ World Meteorological Day	– 23 march
☆ World health Day	– 7 april
☆ World Space Day	– 12 April
☆ World Heritage Day	– 18 April
☆ World Earth Day	– 22 April
☆ National Panchayati Raj Day	– 24 April
☆ World Veterinary Day	– 28 april
☆ International Labour Day	– 1 May
☆ International Biodiversity Day	– 22 May
☆ World No Tobacco Day	– 31 may
☆ World Milk Day	– 1 June
☆ World Environment Day	– 5 June

- ☆ National Agriculture Day — 1 July
- ☆ Van Mahotsava — 1 – 7 July
- ☆ World Population Day — 11 July
- ☆ NABARD Foundation Day — 12 July
- ☆ ICAR Foundation Day — 16 July
- ☆ Bank Nationalization Day — 19 July
- ☆ National Forest Day — 28 july
- ☆ National Women Day — 9 August
- ☆ World Coconut Day — 2 September
- ☆ International Literate Day — 8 September
- ☆ World Ozone Day — 16 September
- ☆ Hindi Day — 14 September
- ☆ CSIR Foundation Day — 26 September
- ☆ World Animal Welfare Day — 4 October
- ☆ National Women Farmers Day — 15 October
- ☆ National Milk Day — 26 November
- ☆ World Food Day — 16 October
- ☆ World AIDS Day — 1 December
- ☆ International Computer Literate Day — 2 December
- ☆ World Pollution Prevention Day — 2 December
- ☆ Bhopal Gas Tragedy — 3 December
- ☆ Agriculture Women Day — 4 December
- ☆ World Soil Day — 5 December
- ☆ National Energy Conservation Day — 14 December
- ☆ World Wild Life Day — 20 December
- ☆ Biological Diversity Day — 29 December
- ☆ Kisan Day/National Farmer's Day — 23 December
- ☆ Jai Kishan Jai Vigyan Week — 23 to 29 December

Important Formula Used in Agricultural

$$\text{Leaf area index (LAI)} = \frac{total\ leaf\ area}{ground\ area}$$

$$\text{Leaf equivalent ratio (LER)} = \frac{sole\ crop}{mixed\ crop}$$

$$\text{Harvesting index per cent} = \frac{economic\ yield\ (grain)}{biological\ yield\ (grain+straw)}$$

$$\text{Sodium absorption ratio (SAR)} = \frac{Na^+}{\sqrt{\dfrac{ca^{+2}+mg^{+2}}{2}}}$$

$$\text{Absolute humidity} = \frac{weight\ of\ water\ vapour}{volume\ of\ air}$$

$$\text{Cropping intensity per cent} = \frac{total\ cropped\ area\ in\ year}{Net\ cultuvated\ area} \times 100$$

Crop rotation intensity per cent =

$$\frac{no.of\ crop\ raised\ in\ a\ crop\ rotation}{duration\ of\ crop\ rotation} \times 100$$

$$\text{Cropping index} = \frac{no.of\ crops\ grown\ per\ year}{cropped\ area} \times 100$$

$$\text{Crop water use efficiency} = \frac{crop\ yield}{evapotranspiration}$$

$$\text{Pore space or porosity of soil per cent} = 100 - \frac{bulk\ density}{particle\ density} \times 100$$

$$\text{Moisture in soil per cent} = \frac{loss\ in\ weight}{oven\ dry\ weight} \times 100$$

$$\text{Real value of seed} = \frac{pure\ seed\ \%\ \times germination\ \%}{100}$$

$$\text{Pure live seed} = \frac{purity\ \%\ \times\ germination\%}{100}$$

$$\text{Particle density of soil} = \frac{weight\ of\ soil\ solids}{volume\ of\ soil\ solids}$$

$$\text{Bulk density of soil} = \frac{\text{Weight of dry soil}}{\text{Volume of soil(solids + pores)}}$$

$$\text{Intelligence Quotient (I.Q.)} = \frac{mental\ age}{chronological\ age} \times 100$$

$$\text{Ginning per cent} = \frac{weight\ of\ lint}{weight\ of\ seed\ cotton} \times 100$$

$$\text{Relative humidity} = \frac{water\ vapour\ present\ in\ the\ air}{water\ vapour\ required\ for\ saturation} \times 100$$

$$\text{Celsius and Fahrenheit relation ship} = \frac{C}{5} = \frac{F-32}{9}$$

Important Measurement

Area	100 m² =1 are or 0.025 acre
	100 are = 1 ha or 2.47 acres
	1 acre = 0.405 ha
	100 ha = 1 km² or 0.386 mile
Length	10 mm = 1 cm or 0.39 inch
	100 cm = 1 m or 39.4 inches
	100 m = 1 km or 0.62 mile
Weight	1000 mg = 1 g
	1000 g = 1 kg or 2.21 pound
	1000 kg = 1 tonne
Water flow	1 Cusec = 28.3 lit of water/sec
	1 Cumec = 1000 lit/sec
	1 ha mm = 1000 lit or 10^3
	1 ha cm = 100000 lit or 10^5
	1 ha m = 10000000 lit or 10^7

ICAR and Research Institute

- ☆ Full form of ICAR — **Indian Council of Agricultural Research**
- ☆ Imperial Agriculture Research Institute (IARI) Bihar in **–1905**
- ☆ IARI was set up in Bihar in 1905 by **– Lord Curzon**
- ☆ Imperial Council of Agriculture Research Institute **– 16 July 1929**
- ☆ IARI shift from Bihar to Delhi in **– 1936**
- ☆ Imperial Council change into Indian Council in **– 1946**
- ☆ ICAR as a autonomous organization **– 1965**
- ☆ ICAR was established based on the recommendations of **– Royal Commission**
- ☆ The President of the ICAR is **– Agriculture Minister**
- ☆ ICAR is **– Registered Cooperative**
- ☆ First transfer of technology (TOT) project of ICAR was **– AICPND**
- ☆ ICAR established a section of extension education at its headquarters in **– 1971**
- ☆ Agriculture Scientist Recruitment Board (ASRB) was set up in the year **– 1973**
- ☆ ASRB was set up the recommendations of the committee headed by **– Gajendra Gadkaer**
- ☆ Chairman of ICAR **– Director General (DG)**

☆ President of ICAR — **Minister of Agriculture**

☆ First President of ICAR — **Habibulla**

☆ First DG of ICAR — **Dr. B.P. Pal**

☆ First Indian director of IARI was — **Dr. B. Vishvanath**

☆ First Deputy Director General of Horticulture, ICAR — **Dr. K.L. Chadha**

☆ First Agriculture Minister of Independent India — **Rafi Ahmed Kidwai**

☆ First All India Co-ordinated Research Project on — **Maize (1957)**

☆ First KVK established at — **Puducherry (1974)**

☆ First State Agriculture University of India — **GBPAU, Pantnagar (1960)**

☆ First National Research Centre for — **Groundnut**

☆ New Director General of ICAR — **Dr. Trilochan Mohapatra**

☆ Chairman of ASRB of ICAR — **Dr. Gurubachan Singh**

☆ Publication of ICAR

The Indian Journal of Agriculture Sciences – English

The Indian Journal of Animal Sciences – English

Indian Farming – English

Indian Horticulture – English

Kheti – Hindi

Krishi Chayanika – Hindi

Phal Phool – Hindi

ICAR Awards and Related Field

☆ **ICAR Young Scientist Award** (two years) – Agriculture and Allied Science.

☆ **Dr. C. Subramaniam Award** (two years) – Outstanding Teacher in Agriculture and Allied Science.

☆ **Rafi Ahmed Kidwai Memorial Award** (two years)- Agriculture and Allied Science.

☆ **ICAR Team Award** (three years) – Multidisciplinary Research Work in Agriculture and Allied Science.

☆ **Fakruddin Ali Ahmed Award** (two years) – Agriculture Research in Tribal Field.

☆ **Jawaharlal Nehru Award** (annual)– Best Ph.D in Agriculture and Allied Science.

☆ **Chaudhary Devi Lal Award** (annual) – Outstanding Performance of AICRP.

☆ **Choudhary Charan Singh Award** (annual) – Excellent in Journlalism in Agriculture Research Development.

☆ **Dr. Rajendra Prasad Purushkar** (two years) – Technical Books in Hindi in Agriculture and Allied Science.

☆ **Jagjiwan Ram Kisan Purushkar** (annual) – Two Innovative Farmers in Agriculture and Allied Science.

☆ **N.G. Ranga Farmer Award** (annual) – Innovative Farmers for Diversified Agriculture Activities.

☆ **Hari Om Ashram Trust Award** (Biannual) – Published Research in Crop Science, Horticulture, and Animal Science.

☆ **Punjab Rao Deshmukh Women Agriculture Scientist Award** (annual)– Women Scientist in ICAR Institute.

☆ **Vasant Rao Naik Award** (annual) Water Conservation and Dryland Farming.

☆ **Sardar Patel Outstanding ICAR Institution Award** (annual) – Best Performance in Agriculture Research and Education given to Two ICAR Institute/One SAU.

☆ **Best Krishi Vigyan Kendra Award** (two years) – Outstanding Contribution made by KVK.

☆ The awards conferred by the publication division, Govt. of India to encourage original creative writing in hindi in various disciplines of mass communication is **Bharatendu Harischandra Award.**

Agricultural Universities in India

State Agricultural Universities (63)

☆ Assam Agricultural University	**– Jorhat (Assam)**
☆ Anand Agricultural University	**– Anand (Gujarat)**
☆ Acharya NG Ranga Agricultural University	**–Hyderabad (Telangana)**
☆ Agriculture University Jodhpur	**– Jodhpur (Raj.)**
☆ Agriculture University Kota	**– Kota (Raj.)**
☆ Birsa Agricultural University	**– Ranchi (Jharkhand)**
☆ Bidhan Chandra Krishi Viswavidyalaya	**– Mohanpur (West Bengal)**
☆ Bihar Agriculture University	**– Bhagalpur (Bihar)**
☆ Central Agricultural University	**– Imphal (Manipur)**
☆ Chaudhary Charan Singh Haryana Agricultural University	**– Hissar (Haryana).**

☆ Chaudhary Sarwan Kumar Himachal Pradesh Krishi Vishvavidhalaya-
Palampur (H.P.)

☆ Chandra Shekar Azad University of Agriculture and Technology – **Kanpur (U.P.)**

☆ Chhattisgarh Kamdhenu Vishwavidyalaya **– Durg (C.G.)**

☆ Dr Yashwant Singh Parmar University of Horticulture and Forestry –
Solan (H.P.)

☆ Dr Balasaheb Sawant Konkan Krishi Vidyapeeth – **Dapoli (Maharashtra)**

☆ Dr Panjabrao Deshmukh Krishi Vidyapeeth **– Akola (Maharashtra)**

☆ Dr YSR Horticulture University **– Tadepalligudem (A.P.)**

☆ Govind Ballabh Pant University of Agriculture and Technology –
Pantnagar (Uttarakhand)

☆ Guru Angad Dev University of Veterinary and Animal Sciences –
Ludhiana (Punjab)

☆ Indira Gandhi Krishi Vishwavidyalaya **– Raipur (C.G)**

☆ Jawaharlal Nehru Krishi Vishwavidyalaya **– Jabalpur (M.P)**

☆ Junagadh Agriculture University **– Junagadh (Gujarat)**

☆ Kerala University of Fisheries and Ocean studies **– Kochi (Kerala)**

☆ Kerala Agricultural University **– Thrissur (Kerala)**

☆ Kerala Veterianary and Animal Sciences University – **Thiruvananthapuram (Kerala)**

☆ Karnataka Veterinary Animal and Fisheries Science University
Nandinagar (Karnataka)

☆ Kamdhenu University **– Gandhi nagar (Gujarat)**

☆ Lala Lajpat Rai University of Veterinary and Animal Sciences – **Hisar (Haryana)**

☆ Manyavar Shri Kanshi Ram ji Agriculture Technology **– Banda (U.P.)**

☆ Maharana Pratap Univ. of Agriculture and Technology – **Udaipur (Rajasthan)**

☆ Maharashtra Animal Science and Fishery University **– Nagpur (Maharashtra)**

☆ Mahatma Phule Krishi Vidyapeeth **– Rahuri (Maharashtra)**

☆ Nanaji Deshmukh Veterinary Sciences University – **Jabalpur (M.P.)**

☆ Navsari Agricultural University **– Navsari (Gujarat)**

☆ Narendra Deva University of Agriculture and Technology – **Faizabad (U.P.)**

☆ Odisha Univ. of Agriculture and Technology – **Bhubaneswar (Odisha)**

☆ Prof. Jayashankar Telangana State Agriculture University – **Rajendra Nagar (Hyderabad)**

☆ Punjab Agricultural University – **Ludhiana Punjab)**

☆ Rajendra Agricultural University – **Samastipur (Bihar)**

☆ Rajmata Vijay Raje Sciendia Agricultural University – **Gwalior (M.P.)**

☆ Rani Laxmi Bai Central Agriculture University – **Jhansi (U.P.)**

☆ Rajasthan University of Veterianary and Animal Sciences – **Bikaner (Rajasthan)**

☆ Sardar Vallabh Bhai Patel Univ of Agriculture and Technology – **Meerut (U.P.)**

☆ Sardar krushinagar-Dantiwada Agricultural University – **Dantiwada (Gujarat)**

☆ Sher-E-Kashmir University of Agricultural Sciences and Technology **Jammu (J&K)**

☆ Sher-E-Kashmir University of Agricultural Sciences and Technology – **Srinagar (J&K)**

☆ Sri Venkateswara Veterinary University – **Chittoor (A.P.)**

☆ Sri Karan Narendra Agriculture University – **Jobner (Rajasthan)**

☆ Swami Keshwanand Rajasthan Agriculture University – **Bikaner (Rajasthan)**

☆ Sri Konda Laxman Telangana State Horticultural University – **Rajendra Nagar (Hyderabad)**

☆ Tamil Nadu Agricultural University – **Coimbatore (Tamil Nadu)**

☆ Tamil Nadu Veterinary and Animal Sciences University – **Chennai (Tamil Nadu)**

☆ Tamil Nadu Fisheries University – **Nagapattinam (T.N.)**

☆ University of Agricultural Sciences – **Dharwad (Karnataka)**

☆ University of Agricultural Sciences – **Banglore (Karnataka)**

☆ University of Horticultural Sciences – **Navanagar (Karnataka)**

☆ University of Agricultural Sciences – **Raichur (Karnataka)**

☆ University of Agricultural Sciences – **Shimoga (Karnataka)**

☆ UP Pandit Deen Dayal Upadhaya Pashu Chikitsa Vigyan Vishwa Vidhyalaya Evam Go Anusandhan Sansthan – **Mathura (U.P.)**

☆ Uttar Banga Krishi Vishwaviddyalaya – **Coach Behar (W.Bengal)**

☆ Uttarakhand University of Horticulture and Forestry – **Garhwal (Uttarakhand)**

☆ Vasantrao Naik Marathwada Agricultural University — **Parbhani (Maharashtra)**

☆ West Bengal University of Animal and Fishery Sciences —**Kolkata (W.Bengal)**

Deemed Universities (4)

☆ Indian Agricultural Research Institute (IARI) — **Pusa (New Delhi)**

☆ Indian Veterinary Research Institute (IVRI) — **Izatnagar (U.P.)**

☆ National Dairy Research Institute (NDRI) — **Karnal (Haryana)**

☆ Central Institute of Fisheries Education (CIFE) – **Mumbai (Maharashtra)**

Central Universities with Agriculture Faculty (7)

☆ Banaras Hindu University — **Varanasi (U.P.)**

☆ Aligarh Muslim University — **Aligarh (U.P.)**

☆ Vishwa Bharti — **Shantiniketan (West Bengal)**

☆ Nagaland University — **Medizipherma (Nagaland)**

☆ Allahabad Agricultural University — **Allahabad (U.P.)**

☆ Central Agriculture University — **Iroisemba (Imphal)**

☆ Rani Lakshami Bai University — **Jhansi (U.P.)**

National ICAR Institute (64)

☆ Indian Institute of Soil Sciences (IISS), **Bhopal**

☆ Central Institute of Agricultural Engineering (CIAE), **Bhopal**

☆ Indian Institute of Sugarcane Research (IISR), **Lucknow**

☆ Indian Institute of Vegetable Research (IIVR), **Varanasi**

☆ Indian Institute of Pulses Research (IIPR), **Kanpur**

☆ Vivekananda Parvatiya Krishi Anusandhan Sansthan (VPKAS), **Almora**

☆ National Institute of High Security Animal Diseases, **Bhopal**

☆ National Biotic Stress Management Institute, **Raipur**

☆ Central Institute of Cotton Research, **Nagpur**

☆ Central Institute of Sub Tropical Horticulture, **Lucknow**

☆ Central Institute for Research on Cattle, **Meerut**

☆ Indian Agricultural Statistics Research Institute, **New Delhi**

☆ Indian Institute of Agricultural Biotechnology, **Ranchi**

☆ Indian Institute of Natural Resins and Gums, **Ranchi**

☆ Central Island Agricultural Research Institute, **Port Blair**

☆ Central Institute of Arid Horticulture, **Bikaner**

- ☆ Central Arid Zone Research Institute, **Jodhpur**
- ☆ Central Institute of Freshwater Aquaculture, **Bhubaneshwar**
- ☆ Central Avian Research Institute, **Izatnagar**
- ☆ Central Institute of Fisheries Technology, **Cochin**
- ☆ Central Inland Fisheries Research Institute, **Barrackpore**
- ☆ National Institute of Research on Jute and Allied Fibre Technology, **Kolkata**
- ☆ Central Institute Brackishwater Aquaculture, **Chennai**
- ☆ Central Tuber Crops Research Institute, **Trivandrum**
- ☆ Research Complex for NEH Region, **Barapani**
- ☆ Central Sheep and Wool Research Institute, **Rajasthan**
- ☆ Indian Grassland and Fodder Research Institute, **Jhansi**
- ☆ ICAR Research Complex for Eastern Region, **Patna**
- ☆ Central Institute of Research on Cotton Technology, **Mumbai**
- ☆ National Institute of Abiotic Stress Management, **Malegaon**
- ☆ Central Coastal Agricultural Research Institute, **Goa**
- ☆ Indian Institute of Spices Research, **Calicut**
- ☆ Central Institute on Post harvest Engineering and Technology, **Ludhiana**
- ☆ Central Institute of Temperate Horticulture, **Srinagar**
- ☆ Central Plantation Crops Research Institute, **Kasargod**
- ☆ Central Institute for Research on Goats, **Makhdoom**
- ☆ Central Marine Fisheries Research Institute, **Kochi**
- ☆ Central Potato Research Institute, **Shimla**
- ☆ Central Research Institute for Jute and Allied Fibres, **Barrackpore**
- ☆ Central Research Institute of Dry land Agriculture, **Hyderabad**
- ☆ National Academy of Agricultural Research and Management, **Hyderabad**
- ☆ Sugarcane Breeding Institute, **Coimbatore**
- ☆ Central Rice Research Institute (CRRI), **Cuttack**
- ☆ Central Soil and Water Conservation Research and Training Institute, **Dehradun**
- ☆ Central Soil Salinity Research Institute, **Karnal**
- ☆ Central Tobacco Research Institute, **Rajahmundry**
- ☆ Indian Institute of Horticultural Research, **Bengaluru**
- ☆ National Institute of Animal Nutrition and Physiology, **Bengaluru**
- ☆ National Institute of Veterinary Epidemiology and Disease Informatics, **Bengaluru**

☆ Central Institute for Research on Buffaloes, **Hissar**

☆ National Centre for Agricultural Economics and Policy Research, **New Delhi**

☆ Central Agro forestry Research Institure, **Jhansi**

☆ Central Citrus Research Institute, **Nagpur**

☆ Indian Institute of Maize Research, **New Delhi**

☆ Indian Institute of Wheat and Barley Research, **Karnal**

☆ Indian Institute of Farming System Research, **Modipuram**

☆ Indian Institute of Millets Research, **Hyderabad**

☆ Indian Institute of Oilseeds Research, **Hyderabad**

☆ Indian Institute of Oil Palm Research, **West Godawari**

☆ Indian Institute of Water Management, **Bhubaneshwar**

☆ Indian Institute of Rice Research, **Hyderabad**

☆ Indian Institute of Seed Research, **Mau**

☆ Central Institute for Women in Agriculture, **Bhubaneshwar**

☆ National Organic Farming Research Institute, **Sikkim**

National Research Centres (15)

☆ National Centre for Integrated Pest Management, **New Delhi**

☆ National Research Centre for Banana, **Trichi**

☆ National Research Centre for Grapes, **Pune**

☆ National Research Centre for Litchi, **Muzaffarpur**

☆ National Research Centre for Pomegranate, **Solapur**

☆ National Research Centre on Camel, **Bikaner**

☆ National Research Centre on Equines, **Hisar**

☆ National Research Centre on Meat, **Hyderabad**

☆ National Research Centre on Mithun, **Nagaland**

☆ National Research Centre on Orchids, **Sikkim**

☆ National Research Centre on Pig, **Guwahati**

☆ National Research Centre on Plant Biotechnology, **New Delhi**

☆ National Research Centre on Seed Spices, **Ajmer**

☆ National Research Centre on Yak, **West Kemang**

☆ National Research Centre on Integrated Farming, **Motihari**

National Bureaux (6)

☆ National Bureau of Soil Survey and Land Use Planning, **Nagpur**

☆ National Bureau of Animal Genetic Resources, **Karnal**

☆ National Bureau of Plant Genetics Resources, **New Delhi**

☆ National Bureau of Agriculturally Important Micro-organisms, **Mau, Uttar Pradesh**

☆ National Bureau of Agriculturally Important Insects, **Bengaluru**

☆ National Bureau of Fish Genetic Resources, **Lucknow**

Directorates/Project Directorates (13)

☆ Directorate of Soybean Research, **Indore**

☆ Directorate of Rapeseed and Mustard Research, **Bharatpur**

☆ Directorate of Mushroom Research, **Solan**

☆ Directorate on Onion and Garlic Research, **Pune**

☆ Directorate of Cashew Research, **Puttur**

☆ Directorate of Groundnut Research, **Junagarh**

☆ Directorate of Weed Science Research, **Jabalpur**

☆ Project Directorate on Foot and Mouth Disease, **Mukteshwar**

☆ Directorate of Poultry Research, **Hyderabad**

☆ Directorate of Knowledge Management in Agriculture (DKMA), **New Delhi**

☆ Directorate of Cold Water Fisheries Research, **Nainital**

☆ Directorate of Medicinal and Aromatic Plants Research, **Anand**

☆ Directorate of Floricultural Research, **Pune**

All India Coordinated Research Projects (61)

☆ AICRP on Nematodes – **New Delhi**

☆ AICRP on Maize – **New Delhi**

☆ AICRP Rice – **Hyderabad**

☆ AICRP on Chickpea – **Kanpur**

☆ AICRP on MULLARP – **Kanpur**

☆ AICRP on Pigeon Pea – **Kanpur**

☆ AICRP on Arid Legumes – **Kanpur**

☆ AICRP on Wheat and Barley Improvement Project – **Karnal**

☆ AICRP Sorghum – **Hyderabad**

☆ AICRP on Pearl Millets – **Jodhpur**

☆ AICRP on Small Millets – **Bangaluru**

☆ AICRP on Sugarcane – **Lucknow**

☆ AICRP on Cotton – **Coimbatore**

- ☆ AICRP on Groundnut – **Junagarh**
- ☆ AICRP on Soybean – **Indore**
- ☆ AICRP on Rapeseed and Mustard – **Bharatpur**
- ☆ AICRP on Sunflower, Safflower, Castor – **Hyderabad**
- ☆ AICRP on Linseed – **Kanpur**
- ☆ AICRP on Sesame and Niger – **Jabalpur**
- ☆ AICRP on IPM and Biocontrol – **Bangaluru**
- ☆ AICRP on Honey Bee Research and Training – **Hisar**
- ☆ AICRP -NSP(Crops) – **Mau**
- ☆ AICRP on Forage Crops – **Jhansi**
- ☆ AICRP on Tropical Fruits – **Bangaluru**
- ☆ AICRP Arid Zone Fruits – **Bikaner**
- ☆ AICRP Mushroom – **Solan**
- ☆ AICRP Vegetables including NSP vegetable – **Varanasi**
- ☆ AICRP Potato – **Shimla**
- ☆ AICRP Tuber Crops – **Thiruvananthapuram**
- ☆ AICRP Palms – **Kasaragod**
- ☆ AICRP Cashew – **Puttur**
- ☆ AICRP Spices – **Calicut**
- ☆ AICRP on Medicinal and Aromatic Plants including Betelvine – **Anand**
- ☆ AICRP on Floriculture – **New Delhi**
- ☆ AICRP in Micro Secondary and Pollutant Elements in Soils and Plants – **Bhopal**
- ☆ AICRP on Soil Test with Crop Response – **Bhopal**
- ☆ AICRP on Long Term Fertilizer Experiments – **Bhopal**
- ☆ AICRP on Salt Affected Soils and Use of Saline Water in Agriculture – **Karnal**
- ☆ AICRP on Water Management Research – **Bhubaneshwar**
- ☆ AICRP on Ground Water Utilisation – **Bhubaneshwar**
- ☆ AICRP Dryland Agriculture – **Hyderabad**
- ☆ AICRP Weed Control – **Jabalpur**
- ☆ AICRP on Agroforestry – **Jhansi**
- ☆ AICRP on Farm Implements and Machinery – **Bhopal**
- ☆ AICRP on Utilization of Animal Energy (UAE) – **Bhopal**
- ☆ AICRP on Plasticulture Engineering and Technologies – **Ludhiana**
- ☆ AICRP on PHT – **Ludhiana**

☆ AICRP on Goat Improvement – **Mathura**

☆ AICRP on Cattle Research – **Meerut**

☆ AICRP on Poultry – **Hyderabad**

☆ AICRP-Pig – **Izzatnagar**

☆ AICRP Foot and Mouth Disease – **Mukteshwar**

☆ AICRP ADMAS – **Bangalore**

☆ AICRP on Home Science – **Bhubaneshwar**

☆ AICRP on Agrometeorology (including Network on Impact adaptation and Vulnerability of Indian Agriculture to Climate Change) – **Hyderabad**

☆ AICRP Integrated Farming System Research(including Network Organic Farming) – **Modipuram**

☆ AICRP on Improvement of Feed Sources and Nutrient Utilisation for raising animal production – **Bangaluru**

☆ AICRP on Renewable Sources of Energy for Agriculture and Agro Based Industries – **Bhopal**

☆ AICRP on Pesticides Residues – **New Delhi**

☆ AICRP Sub Tropical Fruits – **Lucknow**

☆ AICRP on Ergonomics and Safety in Agriculture – **Bhopal**

Network Projects (20)

☆ All India Network Project on Pesticides Residues – **New Delhi**

☆ All India Network Project on Underutilised Crops – **New Delhi**

☆ All India Network Project on Tobacco – **Rajahmundry**

☆ Network Project on White Grubs and Other Soil Arthropods pests – **Durgapura**

☆ Network on Agricultural Acarology – **Bangaluru**

☆ Network on Economic Ornithology – **Hyderabad**

☆ All India Network Project on Rodent Control – **Jodhpur**

☆ All India Network Project on Jute and Allied Fibres – **Barrackpore**

☆ Network project on Improvement of Onion and Garlic – **Pune**

☆ Network Bio-fertilizers – **Bhopal**

☆ Network project on Animal Genetic Resources – **Karnal**

☆ Network Programme on Sheep Improvement – **Avikanagar**

☆ Network Project on Buffaloes Improvement – **Hisar**

☆ Network on Gastro Intestinal Parasitism – **Izatnagar**

☆ Network on Haemorrhagic Septicaemia – **Izatnagar**

☆ Network Programme Blue Tongue Disease – **Izatnagar**

☆ Network Project on Harvest and Post Harvest and Value Addition to Natural Resins and Gums – **Ranchi**

☆ Network Project on R and D Support for Process Upgradation of Indigenous Milk products for industrial application – **Karnal**

☆ Network Project on Conservation of Lac Insect Genetics Resources – **Ranchi**

☆ Network project on Agriculture Bioinformatics and Computational Biology – **New Delhi**

Krishi Vigyan Kendra's in India

States	No. of KVKs
ATARI, Zone I, Ludhiana – 69 KVKs	
Himachal Pradesh	13
Jammu and Kashmir	21
Punjab	22
Uttarakhand	13
ATARI, Zone II, Jodhpur– 61 KVKs	
Delhi	1
Haryana	18
Rajasthan	42
ATARI, Zone III, Kanpur– 69 KVKs	
Uttar Pradesh	69 (Hight)
ATARI, Zone IV, Patna– 63 KVKs	
Bihar	39
Jharkhand	24
ATARI, Zone V, Kolkata– 58 KVKs	
A & N Islands	3
Odisha	33
West Bengal	22
ATARI, Zone VI, Guwahati- 44 KVKs	
Assam	25
Arunachal Pradesh	15
Sikkim	4
ATARI, Zone VII, Barapani– 42 KVKs	
Manipur	9
Meghalaya	7

States	No. of KVKs
Mizoram	8
Nagaland	11
Tripura	7
ATARI, Zone VIII, Pune– 77 KVKs (Hight)	
Maharashtra	45
Gujarat	30
Goa	2
ATARI, Zone IX, Jabalpur– 76 KVKs	
Chattisgarh	25
Madhya Pradesh	51
ATARI, Zone X, Hyderabad– 73 KVKs	
Tamil Nadu	30
Puducherry	3
Andhra Pradesh	24
Telangana	16
ATARI, Zone XI, Bengaluru– 48 KVKs	
Karnataka	33
Kerala	14
Lakshadweep	1
Total KVKs in India	**680**

List of changed name of ICAR institute

S.No.	Old name	New name
1.	Directorate of wheat research (DWR)	Indian institute of wheat and barley research (IIWBR) – Karnal
2.	Directorate of water management (DWM)	Indian institute of water management (IIWM) – Bhubaneswar
3.	Project directorate of farming system research (PDFSR)	Indian institute of farming system research (IIFSR) – Meeruth
4.	Central soil and water conservation research and training institute (CSWCRTI)	Indian institute of soil and water conservation (IISWC) – Dehradun
5.	Directorate of oil palm research (DOPR)	Indian institute of oil palm research (IIOPR) – Andhra Pradesh
6.	National research centre for citrus (NRCC)	Central citrus research institute (CCRI) – Nagpur
7.	National research centre for agro forestry (NRCAF)	Central agro forestry research institute (CAFRI) – Jhansi
8.	Directorate of maize research (DMR)	Indian institute of maize research (IIMR) – New delhi
9.	Central agricultural research institute (CARI)	Central island agricultural research institute (CIARI) – Andman and Nikobar
10.	ICAR research complex for Goa	Central coastal agricultural research institute (CCARI) – Goa
11.	Directorate of sorghum research (DSR)	Indian institute of millets research (IIMR) – Hyderabad
12.	Directorate of rice research (DRR)	Indian institute of rice research (IIRR) – Hyderabad
13.	Directorate of research on women in agriculture (DRWR)	Central institute for women in agriculture (CIWA) - Bhubaneswar

Abbreviations

A3P	Accelerated Pulse Production Change Programme
APEDA	Agriculture and Processed Foods Product Export Development Authority
ASCII	American Standard Code for Information Interchange
ATM	Automatic Teller Machine
ADB	Asian Development Bank
AICRP	All India Co-ordinated Research Project
APAARI	Asia Pacific Association for Agricultural Research Institutions
ARIC	Agricultural Research Information Centre
ASI	Anthropological Survey of India
AHRD	Agricultural Human Resource Development Project
ATMA	Agriculture Technology Management Agency
APC	Agriculture Production Commissioner
APRA	Advance Research Project Agency
ANOVA	Analysis of Variance
ACATS	Agriculture Community Access Telecommunication Services
ATIC	Agriculture Technology Information Centre
ALU	Arithmetic Logic Unit
ACC	Agro Cyber Cafe
ATC	Administrative Reform Commission.
AKIS	Agriculture Knowledge and Information System
AIBP	Accelerated Irrigation Benefit Programme

ATARI	Agricultural Technology Application Research Institute
ARISNET	Agriculture Research Information System Network
AGMARK	Agriculture Produce Grading and Marketing Remark
ACM	Association for Computing Machinery
APDCM	Adoptive Differential Pulse Code Modulation Technique
ADO	Active Data Object
ASP	Active Server Page
AGRIS	International Information System for Agricultural Sciences and Technology
AIFPA	All India Food Preserver`s Association
BPO	Business Process Outsourcing
BSI	Botanical Survey of India
BIS	Bureau of Indian Standards
BIOS	Basic Input Output System
BARC	Bhaba Atomic Research Centre
BASIC	Beginners All Purpose Symbolic Instruction Code
CSIR	Council of Scientific and Industrial Research
CAPART	Council for Advancement of People Action and Rural Technology
CFTRI	Central Food Technological Research Institute
CABI	Centre for Agriculture and Biosciences International
CAS	Conditional Access System
CGPRT	Centre for Research And Development of Coarse Grains, Pulse, Roots and Tuber Crops
CRAFICARD	Committee to Review Arrangements for Institutional Credit for Agriculture and Rural Development
CRIDA	Central Research Institute for Dry Land Agriculture
CSI	Computer Society of India
CPU	Central Processing Unit
COIK	Concept If Known
CBDT	Central Board of Direct Taxes
CU	Control Unit
CD	Compact Disk
CRT	Cathode Ray Tube
CDF	Channel Definition Format
CACP	Commission for Agriculture Cost and Price
CSRE	Crash Scheme for Rural Employment
CPM	Critical Path Method
CRS	Community Radio Station

CONA	Community Oriented Need Assessment
COMDDAP	Computer Manufactures Distributors and Dealers Association of The Philippines
CGAL	Central Grain Analysis Laboratory
CASAS	Commonwealth Association of Scientific Agricultural Societies
CEP	Commission on Environmental Planning
DOS	Disk Operating System
DVD	Digital Virtual Disk.
DSIR	Department of Scientific and Individual Research
DDP	Desert Development Programme
DAESI	Diploma in Agriculture Extension Service for Input Dealers
DELTA	Development Education Leadership Teams
DLCC	District Level Coordination Committee
DPAP	Drought Prone Area Programme
DAC	Department of Agriculture and Co-operation
DDC	District Development Committee
DRDA	District Rural Development Agency
DARE	Department of Agriculture Research and Education
DTH	Direct to Home
DRDO	Defence Research and Development Organisation
DAVP	Directorate of Advertising and Visual Publicity
DOAC	Department of Agriculture and Co-operation
DWCRA	Development of Women and Children in Rural Areas
EAS	Employment Assurance Scheme
EPROM	Erasable Programmable Read Only Memory
EEI	Extension Education Institute
EAS	Employment Assurance Scheme
E Book	Electronic Book
EGS	Employment Guarantee Scheme
EEZ	Exclusive Economic Zone
EEPROM	Electrically Erasable Programmable Read Only Memory
FCI	Food Corporation of India
FICCI	Federation of Indian Chambers of Commerce and Industry
FSI	Forest Survey of India
FAI	Fertilizer Association of India
FRI	Formula Translation Forest Research Institute
FAQ	Frequently Asked Question
FPS	Fair Price Shops

FIAC	Farmers Information and Advisory Centers
FAO	Food and Agriculture Organization
FDI	Foreign Direct Investment
FTP	File Transfer Protocol
FSR	Farming System Research
FTC	Farmer Training Centre
FAST	Fully Automated System for Transport
GATT	General Agreement on Tariffs and Trade
GPS	Global Positioning System
GIS	Geographical Information System
GSI	Geological Survey of India
GAP	Good Agricultural Practices
GLP	Good Laboratory Practices
GMP	Good Management Practices
GDP	Gross Domestic Product
GNP	Gross National Product
GATS	General Agreement on Trade in Services
GSLV	Geo-Synchronous Satellite Launching Vehicle
HTML	Hyper Text Mark up Language
HTTP	Hyper Text Transfer Protocol
HYV	High Yielding Variety
IVLP	Institute Village Linkage Programme
ISNAR	International Service for National Agricultural Research
IFFCO	Indian Farmers Fertilizer Cooperative Limited
IMF	International Monetary Fund
IPR	Intellectual Property Rights
IFCI	Industrial Financial Corporation of India
IDBI	Industrial Development Bank of India
ICICI	Industrial Credit and Investment Corporation of India
IBM	International Business Machine
IGRMS	Indira Gandhi Rashtriya Manav Sangrahalaya
IMA	Indian Military Acedemy
IREDA	Indian Renewable Energy Development Agency
IP	Internet Protocol
ISP	Internet Service Provider
IRC	Internet Relay Chat
ITU	International Telecommunication Union
IIS	Internet Information Services

IRC	Internet Relay Chat
IRDP	Integrated Rural Development Programme
ICRISAT	International Crop Research Institute for Semi –Arid Tropics
ILO	International Labour Organization
IMF	International Monetary Fund
ITC	Indian Tobacco Company
ICT	Information and Communication Technology
ISDN	Integrated Services Digital Network
ICDP	Intensive Cotton Development Programme
IPR	Intellectual Property Right
IFYE	International Farm Youth Exchange
IREP	Integrated Rural Energy Programme
ICAR	Indian Council of Agriculture Research
IAAP	Intensive Agriculture Are Programme
IADP	Intensive Agriculture District Programme
ICRISAT	International Crop Research Institute for Semi-Arid and Tropics
JRSY	Jawahar Rojgar Smiridi Yojana
JPEG	Joint Photographic Expert Group
JRY	Jawahar Rozgaar Yojna
KVK	Krishi Vigyan Kendra
KIRAN	Knowledge Innovation Reposition of Agriculture in North East
KISS	Keep It Short Stupid
KCC	Kisan Call Centre
KRIBHCO	Krishak Bharathi Cooperative Ltd.
LAN	Local Area Network
LLP	Lab to Land Programme
LCD	Liquid Crystal Display
LBS	Lead Bank Scheme
LAMPS	Large Size Adivasi Multipurpose Co-operative Society
LEISA	Low External Input Sustainable Agriculture
MANOVA	Multivariate Analysis of Variance
MFAL	Marginal Farmers and Agriculture Labourers Scheme
MWS	Million Well Scheme
MTC	Model Training Courses
MSP	Minimum Supporting Prices
MPEG	Motion Picture Expert Group.
MAGIC	Multi-Parent Advanced Generation Inter-Cross Population
MICR	Magnetic Ink Character Reader.

MMC	Microsoft Management Council.
MIME	Multipurpose Internet Mail Extension
MFN	Most Favoured Nation
MNC	Multi National Corporation
MMS	Multimedia Message System
MNP	Minimum Need Programme.
MANAGE	National Institute of Agricultural Extension Management
NBSSLUP	National Bureau of Soil Survey and Land Use Planning
NISCAIR	National Institute of Science Communication and Information Resources
NARS	National Agriculture Research System
NFSM	Nation Food Security Mission
NMSA	National Mission for Sustainable Agriculture
NDC	National Development Council
NAARM	National Academy of Agriculture Research Management.
NIESBUD	National Institute of Entrepreneurship and Small Business Development
NAIS	National Agriculture Insurance Scheme
NLM	National Literacy Mission
NAEP	National Agriculture Extension Project
NIAM	National Institute of Agriculture Marketing
NIRD	National Institute of Rural Development
NABARD	National Bank of Agriculture and Rural Development
NAAS	National Academy of Agricultural Sciences
NDDB	National Dairy Development Board
NARP	National Agriculture Research Project
NATP	National Agriculture Technology Project
NAL	National Agricultural Library
NACA	Network of Aquacultural Centers of Asia
NAFED	National Agricultural Cooperative Marketing Federation of India Limited
NCPGR	National Centre for Plant Genomic Research
NDA	National Democratic Alliance
NDA	National Defense Academy
NNRMS	National Natural Resources Management System
NPA	Non Performing Assets
NAPCC	National Action Plan on Climate
NPA	National Policy on Agriculture

NRSA	National Remote Sensing Agency
NWDRA	National Watershed Development for Rainfed Areas
NHM	National Horticulture Mission
NABG	National Agricultural Bio-informatics Grid
ODBC	Open Data Base Connectivity
ODA	Official Development Assistance
OECD	Organization for Economic Corporation and Development
OCR	Optical Character Reader
PIREP	Pilot Intensive Rural Employment Project
PRAI	Planning Research and Action Institute
PROM	Programmable Read Only Memory
PAR	Participatory Action Research
PSLV	Polar Satellite Launching Vehicle
PRA	Participatory Rural Appraisal
PRM	Participatory Research Methods
PUFA	Poly Unsaturated Fatty Acids
PACE	Processor for Aerodynamic Computation and Evaluation
PBR	Plant Breeder Rights
PNR	Passenger Numerical Record
PAL	Phase Alternative Line
PAN	Personal Area Network
PMGSY	Pradhan Mantri Gram Sadak Yojana
PACS	Primary Agriculture Co-operative Credit Society
PERT	Programme Evaluation and Review Technique
PDF	Portable Document Format
PEO	Programme Evaluation Organisation
PMRY	Prime Minister's Rozgar Yojna
PCA	Professional Computer Association
PPVFR	Protection of Plant Varieties Farmer Right
PPP	Public Private Partnership
PIRCOM	Project For Intensification of Regional Research on Cotton, Oil Seed and Millets
PEARL	Package for Effective Administration of Registration Laws
RAAKS	Rapid Appraisal of Agriculture Knowledge Systems
RAVY	Rashtriya Krishi Vikas Yojna
RPC	Research Programme Committee
RTP	Rural Television Project
RTI	Recombinant Inbreed Lines

RAM	Random Access Memory
ROM	Read Only Memory
RFC	Request for Comment
RRA	Rapid Rural Appraisal
RAWE	Rural Agricultural Work Experience
RLEGP	Rural Landless Employment Guarantee Programme
RVP	River Valley Project
RADAR	Radio Detection And Ranging
RASI	Rural Access to Services Through Internet
SARI	Sustainable Access in Rural India
SMTP	Simple Mail Transfer Protocol
SREP	Strategic Research and Extension Plans
SASA	Authority State Agriculture Statistical
SSL	Secure Socket Layer
SITE	Satellite Instructional Television Experiment
SIDCO	Small Industry Development Cooperation
SGRY	Sampooran Grameen Rozghar Yojan
SGSY	Swarn Jayanti Gram Swarojgar Yojna
SFDA	Small Farmers Development Agency
SDRs	Special Drawing Rights
SEWA	Self Employed Women Association
SAMETI	State Agriculture Management and Extension Training Institute
SADP	Specific Area Demonstration Programme
SRI	System of Rice Intensification
SMS	Short Messaging System
SAFTA	South Asian Free Trade Agreement
SARS	Severe Acute Respiratory Syndrome
SIDBI	Small Industrial Development Bank of India
SNF	Solids Not Fat
STAWS	Science and Technology Application for Weaker Section
STEP	Satellite Telecommunication Experiment Project
TNC	Trade Negotiation Committee
TMO	Technology Mission on Oilseeds
TAR	Technical Assessment and Refinement
TRP	Target Rating Point
TPDS	Targeted Public Distribution System
TDU	Technological Dissemination Unit

TIFP	Technology Information Facilitation Programme
TOT	Transfer of Technology
TNA	Training Need Assessment
TARP	Technology Assessment and Refinement Project
TRIPS	Trade Related Intellectual Property Rights
TRIMS	Trade Related Investment Measures
TASMAC	Training and Advanced Studies in Management and Communication
TCP	Transfer Control Protocol
TPS	True Potato Seed
TRYSEM	Training of Rural Youth for Self Employment
TTC	Trainer's Training Centre
URI	Uniform Resource Identifies
URL	Universal Resource Locator
USDA	United States Department of Agriculture
UPA	United Progressive Alliance
UNDP	United Nations Development Programme
USAID	United States Agency for International Development
XML	Exchangeable Mark up Language
WSIS	World Summit on Information Society
WTO	World Trade Organization
WAN	Wide Area Network
WAP	Wireless Amplification Protocol
WLL	Wireless in Local Loop
WMO	World Meteorological Organization
WWW	World Wide Web
ZREAC	Zonal Research and Extension Action Committee

References

Annual Report (2012-13). Department of Animal Husbandry, Dairying and Fisheries, Ministry of Agriculture, Government of India, New Delhi.

Agriculture Situation in India (2015). Directorate of Economics and Statistics, Department of Agriculture and Cooperation, Ministry of Agriculture, New Delhi.

Gupta S.N. (2012). Instant Horticulture, Jain Brother Publication, New Delhi, India.

ICAR (2010). Handbook of Agriculture, Indian Council of Agricultural Research, New Delhi, India.

Indian Horticulture Data Base (2009). Ministry of Agriculture, Government of India, Gurgaon, India.

Katyayan Arun (2006). Fundamental of Agriculture, Kushal Publication, Varanasi, India.

Karhana P.K. *et al.,* (2012). General Agriculture Sciences, Pratibha Publisher, New Delhi, India.

Kantwa, S.R. (2010). Objective Agriculture, New Vishal Publication, New Delhi, India.

Mehrotra Kushal *et al.,* (2012). General Agriculture Question Bank for ICAR Examinations, Kushal Publication and Distribution, Varanasi, India.

Maitry R.S. *et al.,* (2013). General Agriculture for IARI Ph.D Entrance Exam, Sharma Publisher and Distributor, New Delhi, India.

Maitry R.S. *et al.,* (2011). General Agriculture for ICAR JRF Exam, Sharma Publisher and Distributor, New Delhi, India.

Rao, Ramana (2009). General Studies for ARS, Jain Brother Publication, New Delhi, India.

Rathore, Muniraj Singh (2010). General Agriculture for ICAR Exams. Jain Brother, New Delhi, India.

Reddy, S.R. (2004). Principles of Agronomy, Kalyani Publishers, New Delhi, India.

Singh Atul Kumar *et al.*, (2016). Agriculture Extension Explorer, Daya Publishing House, New Delhi.

Sharma, R.K. *et al.*, (2012). *Question Bank for Agricultural Competitions.* Daya Publishing House, New Delhi, India.

Sharma, R.K. *et al.*, (2017). *Agriculture at a Glance.* Daya Publishing House, New Delhi, India.

Singh, Yatender *et al.*, (2010). General Agriculture. New Vishal Publications, New Delhi, India.

Salaria, S.K. (2004). Horticulture at a Glance, Jain Brother Publication, New Delhi, India.

Karhana Pushpendra (2010). General Agriculture for ICAR Exams. New Vishal Publications, New Delhi, India.

Yadav,Vijay Kumar *et al.*, (2012). General Agriculture. Parmar Publisher and Distributers, Dhanbad (Jharkhand), India.

Website of ICAR.